THE SOCIOBIOLOGY OF VISUAL IMAGES

Dedication: *To those I love and who love me. To those who cut me a break when I deserved it and especially when I did not deserve it.*

Acknowledgements: *I want to thank the head of VBW Publishing, Bobby Bernshausen, and cover designer Rebekah L. Sather.*

"The Sociobiology of Visual Images," by Dr. Anthony Napoleon. ISBN 978-1-947532-34-2 (softcover), 978-1-947532-65-6 (hardcover), 978-1-947532-66-3 (eBook).

Published 2018 by Virtualbookworm.com Publishing Inc., P.O. Box 9949, College Station, TX 77842, US.

TABLE OF CONTENTS

PREFACE

A perceived visual image is not merely its inverse impression made upon your retina.

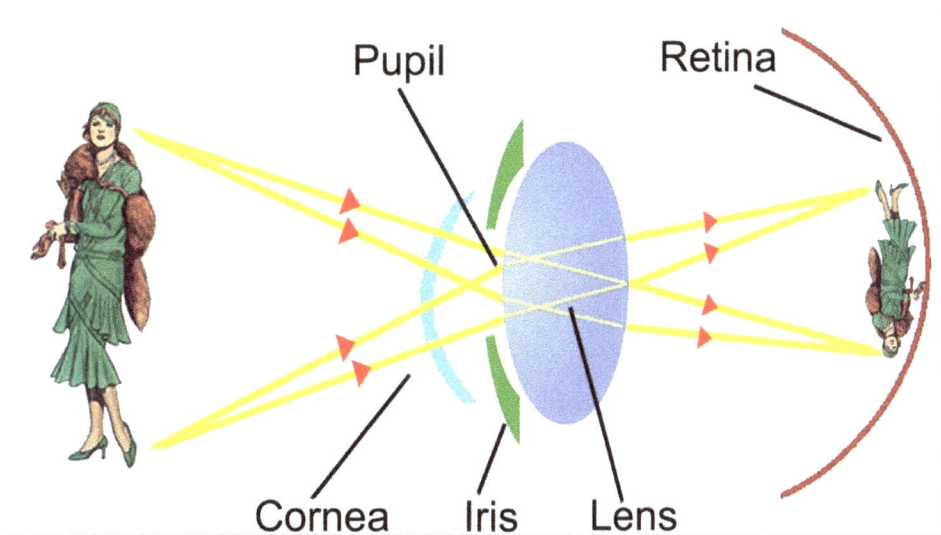

A perceived visual image creates one or more mental images. Once those mental images are experienced by your central nervous system (CNS), they may create a firestorm of feelings, thoughts and spur any number of behaviors.

Once a mental image has been experienced, it is transformed into memory engrams, which take up residence in your brain. Many of these engrams continue to influence your thoughts, feelings and behaviors throughout your life. [1]

Mental images, and the myriad of thoughts, feelings and behaviors associated with them, can be recalled by closing one's eyes and imagining the visual images you've perceived in the past. You can also create a mental image seemingly out of thin air. Created mental images are most often derived from your catalogue of visual images previously experienced.

---

[1] Liu, X., Ramirez, S., Pang, P. T., Puryear, C. B., Govindarajan, A., Deisseroth, K., Tonegawa, S. (22 March 2012). Optogenetic stimulation of a hippocampal engram activates fear memory recall. Nature 22 March 2012 (Vol. 484 Issue. 7394 p. 381-385) DOI: 10.1038/nature11028.

The mental images you create out of thin air can be transformed into visual images that can be shared with other people in the form of drawings, photos, etc. Once a created visual image is shared, it will be transformed by its perceiver into mental images that create feelings, thoughts and resultant behaviors unique to that person.

For example, when Stan Lee created Spiderman, he created a visual image that most of you have perceived and, paradoxically, transformed right back into its genesis form, i.e., a mental image of Spidey similar to the one first experienced by Stan Lee in his mind, but that now incorporates your unique affective and cognitive engrams.

Spiderman

One particular visual image has special importance and meaning to your CNS. That visual image is a face, particularly human faces. This is because faces, both human and animal, produce critically important information necessary for your survival. In fact, faces are so important to human survival, our sensory system automatically scans your environment for them. Our sensory system is so dedicated to finding faces that it is notorious for "seeing" faces where there are no faces to be seen. Seeing faces from random sets of visual stimuli is called Pareidolia. Pareidolia is a type of Apophenia, which is the general tendency of people to see patterns in random sensory stimuli.

The faces people "see" in random patterns always reflect their particular cognitive and cultural programming. Thus, Christians tend to "see" religious figures found in Christianity, Hindus "see" their religious Gods, etc. In early 2000, a Christian woman from South Florida "saw" the Virgin Mary's face in a toasted cheese sandwich. If you have ever wondered what a toasted cheese sandwich emblazoned with the "face" of the Virgin Mary is worth, wonder no more. In 2004, the woman sold her toasted cheese sandwich for $28,000 on eBay.

Example of Pareidolia (Catholic Visual Image of the Virgin Mary or Jean Harlow?)

Jean Harlow

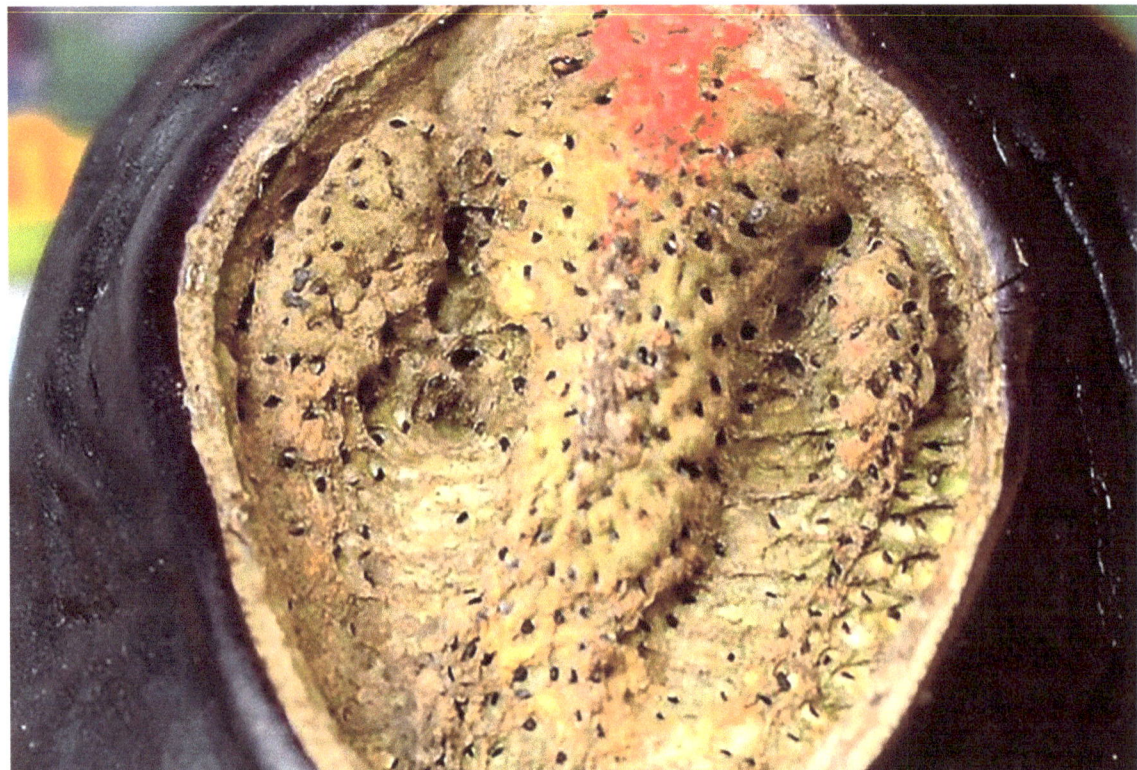

Example of Pareidolia (Hindu God Ganesha)

Ganesha

Right at the outset, I want to address a cognitive psychology problem that research has demonstrated is the reader's enemy when it comes to learning something new or having your awareness broadened from this or any other book.

The problem arises when the reader recognizes a word or term, e.g., "visual image" or "beauty." Once the reader recognizes a word or term, he engages something called "confirmation bias." This is the definition of confirmation bias:

> *"Confirmation bias, also called confirmatory bias or my side bias, is the tendency to search for, interpret, favor, and* **recall information in a way that confirms one's preexisting beliefs or hypotheses**. *It is a type of cognitive bias and* **a systematic error of inductive reasoning**. *People display this bias when they gather or remember information selectively, or when they interpret it in a biased way. The effect is stronger for emotionally charged issues and for deeply entrenched beliefs."* [2]

---

[2] Nickerson, Raymond S. (1998). "Confirmation Bias: A Ubiquitous Phenomenon in Many Guises." Review of General Psychology. 2 (2): 175–220. doi:10.1037/1089-2680.2.2.175.

Short of working one-on-one with each reader in order to disabuse them of their tendencies to engage in confirmation bias, the reader is left to monitor his or her behavior so as to insure that the NEW information, concepts and models of behavior you will find herein are NOT assimilated into the reader's existent understanding of the world.

I suggest to the reader that before she reads any further, she take inventory of her own personal experiences wherein she engaged her confirmation bias tendencies. If you can't find a plethora of examples in your own life, then you may never learn anything new from this or any other book. I'll leave you with this disturbing fact that illustrates how DESTRUCTIVE confirmation bias is when it comes to human development, and may help you to better understand why I am dedicating the beginning of this book to the subject.

By the time you were 5 years old, you had learned approximately 80 percent of the vocabulary you would use for the rest of your life. How can that be? Well, it is not because you were not exposed to and taught new words; in fact, you were exposed to tens of thousands of new words throughout your life but YOU NEVER ADDED THEM TO YOUR VOCABULARY. Why? Because with each new word exposure, you UNCONSCIOUSLY engaged your confirmation bias. In other words, your brain took the new word and then substituted a word you already knew and felt comfortable with in its place. Once that occurred, the new word was discarded and you were left having learned nothing.

Confirmation bias also encompasses prejudices, biases and ideological projections. The very nature of sociobiology "triggers" the biologically ignorant, who are often quick to enforce their uninformed sense of social justice by ascribing any number of stigmatized labels onto anyone who would dare to broach a verboten subject, regardless of how scholarly, fair or unbiased the author. For those readers who can tame their knee-jerk and ideologically driven judgments, this book may help to reveal a world that is more fair and just than so-called "social justice warriors" could ever imagine.

INTRODUCTION

You choose your girlfriends, boyfriends, cars, homes, personal care products, and virtually everything else in your life primarily based upon the quality and dimensions of their visual stimuli you transform into mental images.

Most people have no idea that visual stimuli and their various effects on humans and other animals is one of the most researched areas in all of scientific psychology. This body of research is seldom discussed in the media because editors and producers have concluded that their audience have no interest in anything but the shallow and ethereal aspects of looks, and/or only cover those related topics advertisers pay for.

A lot of people hate contemplating the subject of visual image because it makes them feel uncomfortable. And why does it make them feel uncomfortable? Because it makes them think about a subject that encourages uneasy self-reflection.

At the outset, simply appreciate that visual images, and their resultant mental images, touch us on a deep level. If you are like the average person, you've utilized any number of psychological defenses, which keep your conscious mind from realizing how much of an effect mental and visual images (yours and others) have had on everything you do and has been done to you.

I will now provide the reader a definition of one particular genre of visual image that has a special meaning to human beings. We refer to these visual and mental images as "beauty."

> *Beauty is a characteristic of an animal, idea, object, person or place that provides a perceptual experience of pleasure or satisfaction. Beauty is studied as part of aesthetics, culture, social psychology, philosophy and sociology. An "ideal beauty" is an entity, which is admired, or possesses features widely attributed to beauty in a particular culture, for perfection.* [3]

Our brief comments up to this point will serve as the reader's introduction to the fascinating world of THE SOCIOBIOLOGY OF VISUAL IMAGES, a subject for which I

---

[3] Wikipedia.

have a special expertise and interest. We are now ready to enter the multilayered world of visual and mental images in all of their glory, fascination and profundity.

>*"Sociobiology is a field of scientific study that is based on the hypothesis that social behavior has resulted from evolution and attempts to examine and explain social behavior within that context. It is a branch of biology that deals with social behavior, and also draws from ethology, anthropology, evolution, zoology, archaeology, population genetics, and other disciplines. Within the study of human societies, sociobiology is closely allied to Darwinian anthropology, human behavioral ecology and evolutionary psychology.*
>
>*Sociobiology investigates social behaviors, such as mating patterns, territorial fights, pack hunting, and the hive society of social insects. It argues that just as selection pressure led to animals evolving useful ways of interacting with the natural environment, it led to the genetic evolution of advantageous social behavior."* [4]

Let's begin to discuss the formal subject matter of this book with a non-threatening and easy topic. At least, it will seem that way at first.

BEAUTY AND CARS

Did you ever notice that luxury automobile advertisements that target men don't focus so much on the engineering features of the car, as much as they do the *promised* mental images of the drivers who buy them? Of all the promised experiences sold to the prospective male new car buyer, none is as enticing as the mental image of becoming desired by beautiful women. Here are a couple of representative examples that illustrate this fact.

In the following 2015 Cadillac television commercial, we see a rather nerdy-looking fellow walking down a city street. As he walks along, the camera focuses upon a number of model-like beauties that seemingly stare at the man in a seductive way. At first the man doesn't notice the women's gaze, but when he does, he changes his gait to reflect his newfound confidence.

---

[4] Wilson, Edward O. (1975). Sociobiology, The Abridged Edition. The Belknap Press of Harvard University, Cambridge, Massachusetts.

Cadillac TV Commercial Stills 2015

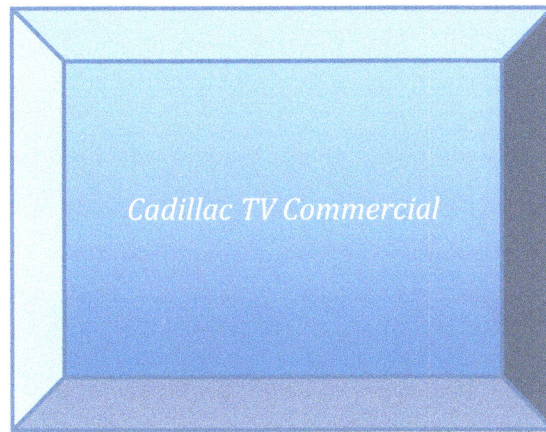

https://youtu.be/aMUZNbyYU8M

What the viewer of this advertisement realizes, at some point, is the beautiful women in the ad are not looking at the man, but instead are looking at a new, shiny Cadillac. The awkward-looking man is merely in the beautiful women's line of sight, which is seductively fixated upon the Cadillac.

If you have ever attended an auto show, you can attest to the fact that beautiful models and new cars go together like mono and diglycerides. The not-so-subliminal messages communicated in the preceding Cadillac commercial and auto shows is that new shiny cars and beautiful women go together, and if you buy one (the car) you can have the other (the babe).

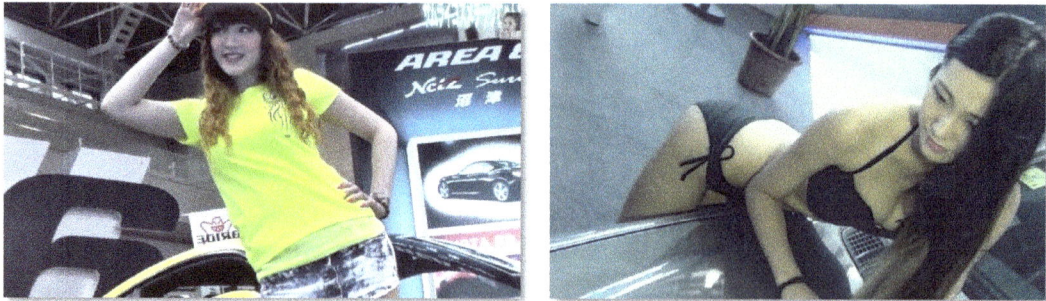

Representative Auto Show Models and Their Poses

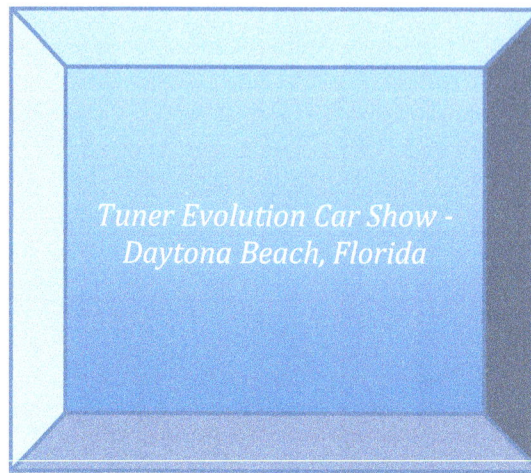

Tuner Evolution Car Show - Daytona Beach, Florida

https://youtu.be/nQifN63uufg

You have to ask yourself, are advertisers selling visual images of physical objects we label "cars," or are they selling mental images that promise men that they can acquire attention and, perhaps, sex from beautiful women? The more interesting question from a clinical psychologist's point of view is this: Do women really go for the guy who drives an expensive new car? Madison Avenue has been paid hundreds of millions of dollars over the decades convincing automobile manufacturers that they know the answer to that question.

One car ad, that may help those of you who think you know more than Madison Avenue and billion-dollar auto manufacturers, utilized a focus group to compare visual images of an entry-level sedan with a Chevrolet truck. The focus group used in this TV ad was comprised of people from all walks of life, ages and genders, with one group exclusively comprised of young and attractive single women.

The women in the focus group were shown two visual images and then asked which of the two men pictured was more attractive, the guy pictured with the truck or the guy pictured with the entry-level sedan. One of the more interesting aspects of this focus group is that the same man had been Photoshopped into two different photos, one showing him standing in front of an entry-level sedan and the other next to a Chevrolet truck. When the women were asked, "Which of the two men is sexier?", the young women in the focus group IMMEDIATELY pointed to the visual image of the man standing in front of the truck, even though **it was the same man in both photos.** By the way, the man was dressed and posed exactly the same in both photos. Backdrops and lighting were, likewise, identical in both photos.

Chevy Truck Commercial Stills Showing the Same Man with Different Vehicles
"Which man is sexier, the man on the left or the man on the right?

Young woman IMMEDIATELY points to the man pictured with the truck as the sexier of the two

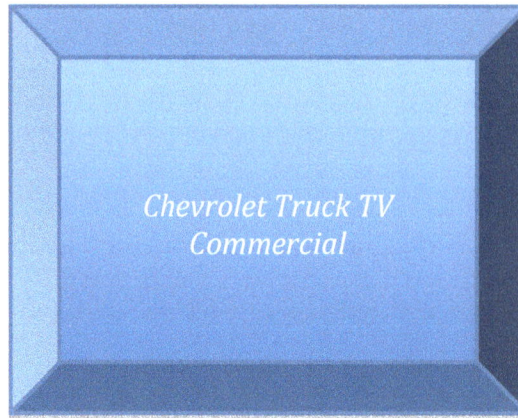

Chevrolet Truck TV Commercial

https://youtu.be/ikH948cqoxw

There are informal studies that have examined the behavior of women in response to a male stranger who asked them to watch his luxury or average-looking car for just five minutes, while he purportedly took care of a pressing matter that couldn't wait. In the first scenario, the man points out a new sports car to the unwitting female subject. In the control scenario, he points out an older, non-descript average car. Want to guess which visual image made the woman want to help the guy?

Not only did the women in this informal study offer to help the man driving the new luxury sports car significantly more often, when compared to the man driving the older entry-level car, many of the women in this study made an overt sexual pass at the guy who was driving the expensive car.

In yet another of these makeshift studies, a young man is seen approaching a woman walking with her boyfriend. He asks her if she would like to go for a short ride around the block with him. She refuses at first, until he points out his shiny new Lamborghini. The woman queries the Lamborghini driver with questions like, "That's not your car, is it?" At which point the driver opens the Lamborghini's door and invites the woman to take a ride with him. Some of the women leave their shell-shocked boyfriends standing alone on the sidewalk, telling them, "We're just going around the block."

A stranger approaches a woman and asks her to take a ride around the block with him

Woman getting into a stranger's Lamborghini

Lamborghini and Stranger Video

https://youtu.be/q43YECOHrEU

A variation of this informal experiment depicts a man who approaches a woman walking to her parked car. The man walks up and says words to the effect, "Hey, I've been watching you, I think you're very pretty. Would you like to go to dinner?" The woman continues to walk away, turning her back on the man as she tells him, "I don't go out to dinner." The man then walks over to his new Lamborghini and opens the door to get in, but before he does, he says to the woman, "I don't generally ask twice." What happens next is fascinating because the woman stops dead in her tracks, turns around and walks back toward the man and says, "Perhaps we could go to a bar. I know a nice bar just down the street." The fellow at that point tells the woman, "I don't go out with gold diggers."

So what is it about having a new luxury car that appears to alter a woman's receptivity to a stranger, underscore STRANGER? Is it as simple as women are materialistic? Not really. Put aside for a moment the cognitive error on the part of the women in these informal studies that fails to realize the vast majority of luxury cars are leased or can be rented. What the women in these makeshift studies are doing is responding to a visual image SYMBOL of a successful male. Financial success triggers both conscious and unconscious sexual receptivity mental images in her CNS.

One thing a display (visual images) of financial success suggests to her conscious mind (mental images) is that this man driving this expensive car can afford to free her from the perils and pitfalls of financial hardship and/or provide her with a standard of living she dreams about. But that is only part of the equation. The unconscious attraction modulator is that the man driving the new luxury car *may be* a sociobiologically viable male, i.e., his success as a male/mating choice *may have* been proven and/or corroborated by the presence (visual image) of the expensive car.

Male genetic viability to a woman means this to her *unconscious* mind: "This male has run the gauntlet of competition on a measure of how well he has gathered resources as evidenced by the expensive car." Displays of material success means to a woman's limbic CNS centers: "This man can take better care of my genes and my offspring." This is what is going on in many women's CNS limbic centers. Keep in mind that the unconscious mind doesn't trade in words and sentences, but feelings and vague impressions created in the first place by mental images. In other words, the vast majority of women could never, even if they desired, put into words what is going on in their limbic systems.

Older and healthy men who display the vestiges of material success tend to be more attractive to women of all ages than are younger men who display the vestiges of material success. Looking through a sociobiology prism, a wealthy and healthy older man has, by virtue of his age, proven that he has run the natural selection gauntlet (life is a marathon, not a 100-meter dash) and is, therefore, a known successful commodity.

Younger men may be posers and/or may be trust fund boys who only look the part. Keep in mind that my description here is NOT what goes on in women's consciousness. It is much more primitive and takes the form of feelings and images that affect her limbic system.

The conveyance of thoughts about sociobiology oftentimes is conflated with the conscious processing of thoughts and ideas. The fact that sociobiology originates in areas of a person's brain that are alienated from cortical thought is a point that must be factored into any true discernment of the influence and power of sociobiology. To say it another way, debating the credibility of sociobiology by relying upon cortical reasoning is like denying the existence of arachnophobia by arguing that statistically, it makes no cortical reasoning sense.

I've interviewed several particularly insightful women who recognized that, for whatever reason or reasons, they were sexually receptive to men who displayed the vestiges of material success. What the more thoughtful women tell me is that a healthy-looking man who drives a new luxury car, for example, can move that man from the "ignore" or "deselected" categories into the "maybe" category. The first date, these women tell me, is reserved to further screen the man to see if he deserves to be moved from the "maybe" category into the "selected" category.

If it sounds like beautiful women think they hold all they cards, they believe that to be the case. But is that true? Do things work out for beautiful women who pre-select the man who drives a new luxury car?

Consider these facts: studies show that men quickly tire of this genre of woman, i.e., the attractive woman who pre-selects men based upon their displays of material success. Beauty is a commodity. The vast majority of men who are aware that they are attractive to attractive women find this genre of woman to be a fungible commodity and a depreciating asset.

As a depreciating asset, beauty and its owner lose their value over time. The time frame window, during which the asset of beauty is marketable, is deceptively short, a fact unknown and/or denied by the vast majority of women who possess the commodity of beauty.

> ***Beautiful women think that men are interested in them (the person) when in reality men are interested in "it," with "it" being her visual stimuli we label "beauty."***

It is beauty that is a depreciating asset, and like all depreciating assets, including new luxury cars, its value will be less a year from now when compared to today. So let me interject and address what some of you egalitarians are undoubtedly thinking. "Why isn't that true for men as well as women?"

As I previously discussed, an older healthy man who is also financially successful is a proven commodity and is, therefore, more valuable than a depreciating commodity. Biology for a man means that a healthy 70-year-old can still father a child; whereas a 70 year-old woman is not only sterile, but also no longer in the same sociobiological league as her 20-something "competitors."

The differences between male and female age-related changes to the face, for example, reflect these sociobiological gender disparities. These differences are seen clearly in the different surgical considerations that are made when performing a male versus female facelift. Let me quote from San Francisco plastic surgeon Dr. Timothy Marten. Dr. Marten, who has authored over 20 textbook chapters on facelift surgery and related topics, presented "Facelift for the male patient—is there really a difference?" during the 2017 Cosmetic Surgery and Aesthetic Dermatology meeting in Las Vegas.

> *"... [P]lastic surgeons have recognized that a diminished margin for error arguably exists in male patients, due to the fact that men have fewer options to conceal their scars or a mediocre result," Dr. Marten says. "Plastic surgeons have also come to recognize that, in most*

cultures, male facial aesthetics differ from those thought desirable in females and attractive masculinity is not as closely equated with youth and beauty as is femininity. As a result, men generally seek a somewhat different outcome from facial rejuvenation surgery, and this has led to a rethinking of techniques that have evolved mainly to treat facial aging in women." New approaches and a rethinking of aesthetic goals now allow for an effective rejuvenation of the male face, while preserving a natural, masculine appearance, he says.

*It's not appropriate to arbitrarily apply concepts and techniques that evolved for treating the female face to that of a man, according to Dr. Marten. "The male face is arguably more nuanced than that of a woman, whom we have come to accept as having a more contrived and made-up appearance. **And certain aspects of male facial aging are often regarded as signs of experience, wisdom and power that many men wish not to lose,"** he says.* [5]

Let me give you just one of thousands of examples of how beauty is a depreciating asset when it comes to women. In the 1960s, a blond bombshell by the name of Brigitte Bardot was considered to be a stunningly beautiful ingénue. Her beauty, in large measure, purchased fame and fortune. What follows are two photos of Brigitte Bardot. The first photo is Brigitte before depreciation of her asset of beauty, followed by a photo of her post-depreciation:

---

[5] Cosmetic Surgery Times: *The Male Facelift Patient is Different.* By Lisette Hilton, June 9, 2017.

A Younger Brigitte Bardot (pre-depreciation)

An Older Brigitte Bardot (post-depreciation)

Miss Bardot has gone on to do marvelous work on behalf of animal protection and welfare.

Men—and I'm especially referencing those men who play the game of displaying their wealth by the cars they drive for the purpose of attracting young beautiful women—seldom, if ever, invest in a depreciating asset for the long term. They lease it, in the same way they lease a car, and then turn it in at the end of the lease for a new model.

Men who invest for the long haul, on the other hand, tend to be interested in classic cars and classy women. And keeping with the analogy, these men make long-term investments in cars because of their engineering, their handling and their investment value.

Before all the depreciating assets and car buyers out there get their hopes up, realize that ONLY a small fraction of car drivers are collectors of the classics (because they cannot afford a classic) and only a small percentage of cars are ever going to be classic.

This leaves us with the situation we saw illustrated in the examples provided thus far: the mating game for the masses is one of subterfuge using visual images, where men engaged in this subterfuge drive a car they really can't afford in order to fool the unsuspecting depreciating asset who is not focused upon character. This brings us to yet another fascinating question: Would it be better for a rich man to NOT drive a new luxury car when it comes to the dating game?

It depends. If a rich man **only** wants a sexual liaison, then attracting women by driving a new luxury car is a very good strategy. Driving an expensive car is a great way to accomplish the "love 'em and leave 'em" gambit. On the other hand, if you are a rich man who wants to screen out women looking for a financier or sugar daddy, and/or just a temporary plaything, then driving around in a quarter of a million dollar car is a huge mistake. Of course, you'll never see that insight in a luxury car advertisement, will you?

PUT ON A HAPPY FACE

The stage play Bye Bye Birdie included a song entitled "Put on a Happy Face." Here are the lyrics from the beginning of that song:

Gray skies are gonna clear up,

Put on a happy face;

Brush off the clouds and cheer up,

Put on a happy face.

Take off the gloomy mask of tragedy,

It's not your style;

You'll look so good that you'll be glad

Ya' decide to smile!

Pick out a pleasant outlook,

Stick out that noble chin;

Wipe off that "full of doubt" look,

Slap on a happy grin!

And spread sunshine all over the place,

Just put on a happy face!

Humans have 46 muscles in their face. Combinations of those 46 muscles contracting and relaxing in various patterns create the expressions that tell others about what *may* be going on inside of us. I use the word "may" because humans make attempts to **fake** various facial configurations in order to gain advantage over their fellow human beings. Moreover, most people are not quite as astute at interpreting facial displays of emotion as they think they are.

Read the opening lyrics to the song "Smiling Faces Sometimes." This song warns the listener about always trusting the facial configuration we call a smile:

Smiling faces sometimes pretend to be your friend

Smiling faces show no traces of the evil that lurks within

Smiling faces, smiling faces sometimes

They don't tell the truth uh

Smiling faces, smiling faces

Tell lies and I got proof

The truth is in the eyes

Cause the eyes don't lie, amen

*Remember **a smile is just***

A frown turned upside down

My friend let me tell you

Smiling faces, smiling faces sometimes

They don't tell the truth, uh

Smiling faces, smiling faces

Tell lies and I got proof

"Smiling Faces Sometimes"
Performed by: The Undisputed
Truth

https://youtu.be/BUmNud_3lcQ

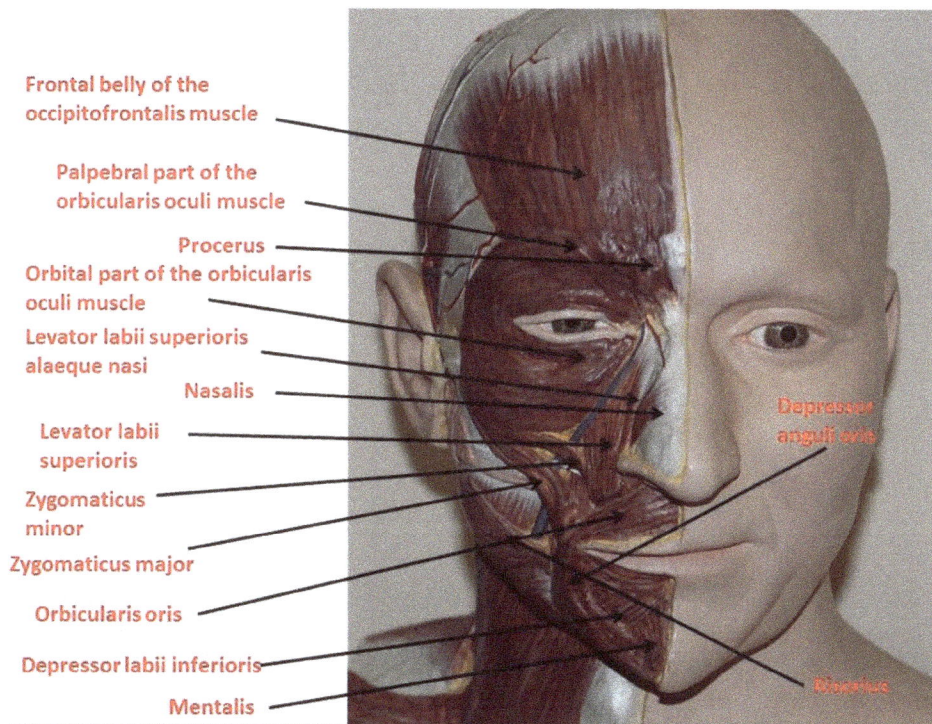

Frontal belly of the
occipitofrontalis muscle

Palpebral part of the
orbicularis oculi muscle

Procerus

Orbital part of the orbicularis
oculi muscle

Levator labii superioris
alaeque nasi

Nasalis

Levator labii
superioris

Zygomaticus
minor

Zygomaticus major

Orbicularis oris

Depressor labii inferioris

Mentalis

Depressor
anguli oris

Risorius

Facial Musculature

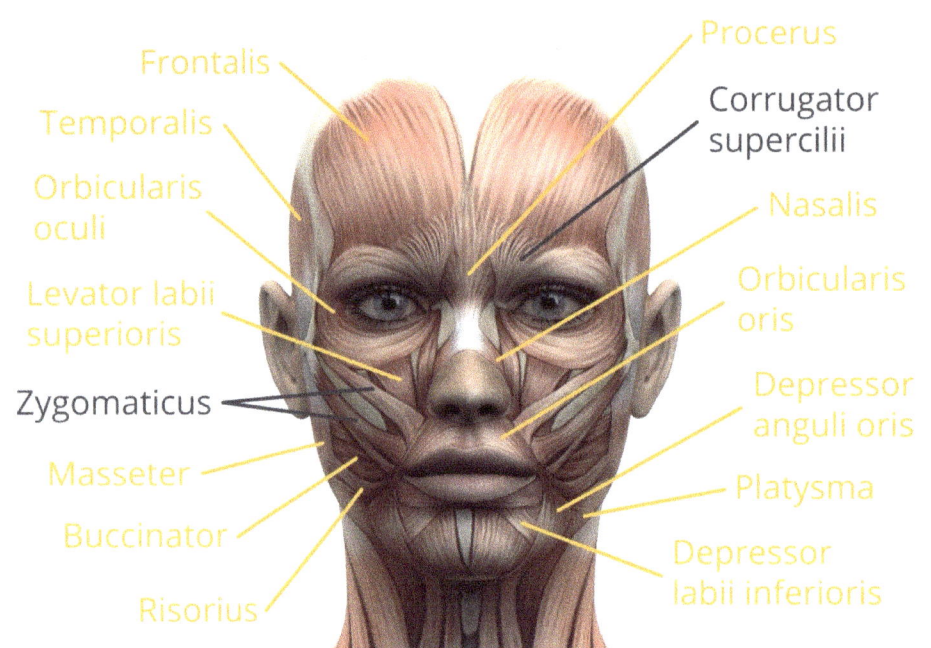

Facial Musculature

I have found that you have to be an exceptional actor to **effectively** fake a smile. Most attempts at a fake smile look disingenuous and are, in fact, worse than no smile at all. People are hardwired to prefer genuine smiles. Baby humans, just a few hours old, long before any learning can take place, respond with a genuine smile in response to visual images of faces depicting a smile, and often cry when shown a picture of a frown.

The French anatomist Guillaume Duchenne was one of the first people to scientifically analyze genuine versus fake smiles. In 1862 he wrote a book entitled *Mecanisme de la Physionomie Humaine*. Duchenne identified for the first time that the zygomatic major muscles (the muscles that when activated pull up the corners of the mouth) can be willed (faked) into action. Duchenne took special note of the genuine smile when he wrote:

> *"Only the sweet emotions of the soul force the orbicularis oculi to contract. Its inertia, in smiling, unmasks a false friend."* [6]

---

[6] Duchenne, G.B. (1862). Mécanisme de la physionomie humaine, ou analyse électro-physiologique de ses différents modes de l'expression. Paris: Archives générales de médecine, P. Asselin.

The fact that facial expressions may reflect the inner emotional state of humans has been known for a very long time. The nuanced meaning that can be communicated by the human face is breathtaking for its depth, breadth and sheer number of emotions, thoughts and states of mind that it can display. Let's now parse this subject of non-verbal communication and facial configurations.

A term like "bedroom eyes" is merely a label for a particular facial configuration involving the eyes and surrounding areas. The same can be said for the labels of "arrogance," "sadness," "anxiety," "seething anger," "disingenuousness" and an almost infinite number of other subtle feelings and states of mind denoted by labels that are communicated by visual images we refer to as facial expressions. What most people don't realize, however, is that not only do internal emotional and cognitive states *produce* facial configurations that may reflect a person's internal environment, but also, facial expressions *can alter* your emotional and cognitive functioning. [7]

Studies have shown that smiling tends to make a person happier. Conversely, frowning tends to make a person more critical. Frowning while reading a troubling narrative makes a person *better able to comprehend* what they are reading. In fact, studies have looked at people who have had Botox™ injections in order to relax their corrugator supercilii muscles. These are the muscles that create the lines in the brow area between the eyes that people refer to as frown lines (glabellar frown lines). Consider this: what these studies have documented is that compassionate people are more likely to be better discerners of others' emotional states.

People who have had Botox™ injections can no longer frown as much, or sometimes not at all. **They also have more difficulty *comprehending* emotionally troubling written narratives that convey sadness and anger.** You

---

[7] c.f. Buck, Ross (1980). "Nonverbal Behavior and the Theory of Emotion: The Facial Feedback Hypothesis". Journal of Personality and Social Psychology. 38 (5): 813.

see, our brains use our facial expressions to help it comprehend parallel emotional states as conveyed by others and through the written word. [8]

The idea is that our ability to empathize with another person's pain or joy is intertwined with our own parallel emotional response to the other person that is reinforced by the genuine display of genuine empathy on our face. The reader should think about that truth next time he or she conveys a personally heart-wrenching life experience to a person who listens to your story with a flat facial affect. Even worse than a flat affect, imagine someone who listens to your tale of woe with a phony facial expression that is fabricated to convey, e.g., "I feel your pain."

My colleagues in the movie business tell me that performers who have made extensive use of Botox™ are not as good at displaying emotions as they once were. In as much that a significant part of the art of acting is reacting, all 46 facial muscles have to be in tip-top working order if the actor is going to be able to genuinely react to his fellow actors' facial expressions. Actions and reactions that are genuine are the secret to displaying accurate and believable facial expressions, whether you are an actor, a doctor, a lawyer, or work in almost any other profession. So go ahead and put on a happy face, as the song says, but that smile has to be genuine. Otherwise people will figure out that your smile is just a frown turned upside down.

WHAT IS BEAUTIFUL IS GOOD - UNLESS IT'S NOT

One of the most famous scientific studies on beauty is titled: *What is Beautiful is Good*. In that study, social psychologists Dion, Berscheid and Walters postulated that physically attractive people have a tremendous advantage over less attractive people that begins from day one and extends until beauty vanishes. [9]

Numerous studies have demonstrated that beautiful women are more likely to be hired, more likely to be the girlfriend/wife a wealthy man, and less likely to be

---

[8] Havas, David A., Glenberg, Arthur M., Gutowski, Karol A. (2010). "Cosmetic Use of Botulinum Toxin-A Affects Processing of Emotional Language." Psychological Science, Volume: 21 issue: 7, page(s): 895-900. June 14.
[9] Dion, K.K., Berscheid, Ellen and Walster, E. (1972). "What is beautiful is good." Journal of Personality and Social Psychology 24(3):285-290, December.

convicted of a crime (depending upon the crime), along with any number of social, vocational and interpersonal advantages. Given this body of research, is it any wonder that my book, *An Illustrated Look at Mankind's Love and **Hatred** of Beauty,* came as a revelation to many people? The love part of the equation is well known, but many people wondered, where did the hatred part of this story come from? Shall we go ahead and pull back even more layers of the onion?

Beauty is highly valued everywhere on planet earth. Beauty is a socially desirable characteristic that gives its owner an advantage over others. Beauty is also a characteristic that is displayed for all to see. Compared to, for instance, wealth and intelligence, which can be hidden (at least in the short run) so as to not engender envy and jealousy in others. Beauty, on the other hand, by its very nature as a visual image, exists in the public domain.

> *Not everyone is beautiful, but every human being is capable of experiencing jealousy and envy when our fellow man has something we want, e.g. beauty.*

Beauty and the privileges that flow from it serve as a constant reminder to average and/or unattractive people that they are denied the benefits that seem to be the birthright of attractive people. This envy effect is so powerful that simply being in the presence of a beautiful woman or handsome man can make average-looking people feel bad. Given that a fundamental precept in social psychology is that we like people who make us feel good about ourselves, imagine how we feel when we are in the presence of someone whose mere existence makes us feel bad about ourselves?

A beautiful girl's popularity begins early in life and conditions her to feel entitled to all of the benefits heaped upon them by little boys and other little girls who want to share in her popularity. But there are some problems that always accompany beauty.

Beautiful women tend to make two fundamental errors about the source of their popularity, an error that often handicaps their character development. Beautiful women assume that the positive attention they receive is because of who they are as a person, when in reality it has little or nothing to do with them, but with a *characteristic* they possess. Let me repeat: men are looking at "it," not at them. It is beauty that people are attracted to, not the person who possesses it.

The second fundamental mistake made by beautiful women is that they fail to realize that *they merely rent beauty, and never own it.* This is because beauty is a fleeting characteristic that fades with the passage of time. Not only does it fade with the passage of time, but also there is a wellspring of new beauty, rented by younger women, continually coming online. When it comes to beauty, it is as if some mysterious economic force maintains the waning beauty of older women and the developing beauty of younger women in perfect equilibrium.

In consideration of beauty's almost perfect economic equilibrium, plastic and cosmetic surgery, as well as every other beauty protecting and replenishing products and services, exists to help economically balance out the loss of beauty due to the passage of time.

In my textbook *Awakening Beauty: An Illustrated Look at Mankind's Love and Hatred of Beauty*, I included a chapter entitled "Murderous Rage." I decided to write that chapter while working as a medical psychology fellow in a plastic surgery surgical center located in La Jolla, CA.

Like almost all plastic surgery practices in Southern California, many of our patients worked in image intensive businesses. We had patients who worked as television journalists, strippers, actors, models and politicians, along with average people who wanted to improve their looks. A number of our more beautiful female patients were referred to our reconstructive practice because someone had disfigured their faces in an act of rage. We saw cuts, scratches, bruises, and even instances where caustic substances had been thrown in our beautiful patients' faces.

I'd like to now mention the forensic case of model Katie Piper. Katie was a beautiful English presenter and model who had her life changed forever when her jilted boyfriend, along with an accomplice, decided to get back at Katie by throwing sulfuric acid in her face.

Surveillance Video Stills from the Attack on Katie Piper. March 31, 2008.

Model Katie Piper Before the Attack

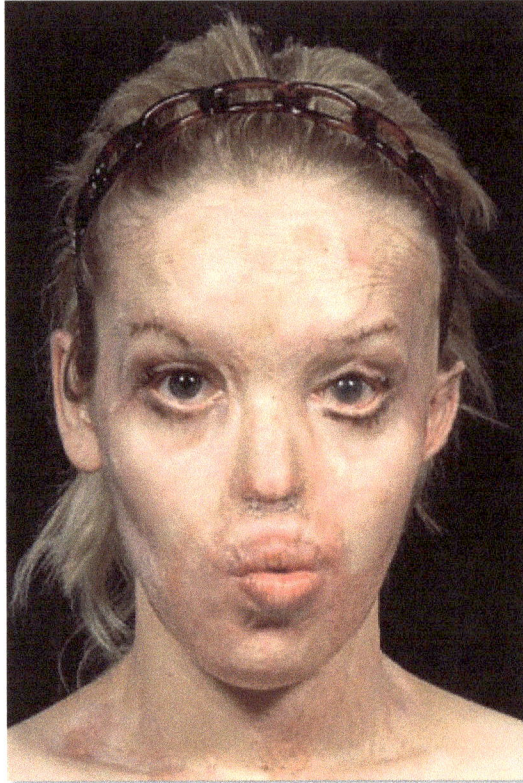

Katie Piper After the Attack

When love morphs into hate, that hatred often focuses upon the characteristic that attracted the attacker to his victim in the first place. And what attracts most men to most women is attractiveness; in particular, a beautiful face.

Don't think for one moment that it is just men who lash out at beauty. One of my early forensic cases involved a beautiful little 10-year-old blond girl who had her hair and face spray-painted with black metal paint. The painter and her co-conspirators held down the little girl while the ringleader spray-painted her victim. Luckily, the little blond girl had her eyes closed and held her breath during the attack.

I tracked down the little girl's attackers in juvenile hall. When I first laid eyes on them, nothing jumped out at me except that they were non-descript looking little girls sitting on a wooden bench at the lockup. These non-descript looking girls, black hair and brown eyes, between the ages of 10 and 13, had ganged up on their victim and, in a primitive act of jealous rage, attacked the girl who had what they didn't have — blue eyes, blond hair and a beautiful face. The little blond girl, described by her teachers as kind and helpful, had committed a sin for which she had to be punished. She had been born beautiful.

What if I told you that some parents become so enraged over their children's beauty that they physically harm them? Read this headline:

Twisted Dad Doused His Three-Year-Old Daughter In Petrol And Burned Her 'Because She Was Too Beautiful'

"A dad-of-three who doused his daughter in petrol and then set fire to her says he did it because his daughter 'was too beautiful', a court has heard.

He then set the girl on fire and poured petrol over his seven-year-old daughter. His younger daughter was left with burns to her face and upper body and required surgery and laser scar management.

*The court heard how around midnight, Herbert told his wife 'the werewolf is coming at 12pm' and 'that's it b****, I am going to kill you' to his daughter before getting petrol and entering her room.*
He then poured petrol over the young child and set her alight before dousing the older child in petrol.

According to reports, the family's neighbor Daniel McMillan arrived shortly after and found Mr. Herbert in the kitchen drinking beer. Herbert then allegedly told Mr. McMillan that they were his kids so he could do whatever he wanted and that his young daughter was too beautiful and that is why he burnt her.

He is accused of two counts of doing an act with intent to unlawfully kill, threatening to kill his partner, being armed with a knife in circumstances likely to cause fear and doing an act likely to endanger life with the intent to harm. Herbert admits burning his child but claims he was insane at the time and so not guilty." [10]

---

[10] Metro UK (2017). Twisted Dad Doused His Three-Year-Old Daughter In Petrol And Burned Her 'Because She Was Too Beautiful. Story by: Tanveer Mann.

Beautiful women know what I'm talking about when it comes to the jealousy and hatred they receive from others. Sexual attraction and beauty are inextricably linked. Sometimes sexual attraction is taboo, if not illegal, when it involves a family member and/or a person under the legal age of consent. One way to do away with forbidden sexual attraction (see the immediately preceding case example) is to kill beauty, i.e., the source of the sexual attraction.

Without exception, every seminar or lecture I've given on this subject found the most beautiful women in the audience IMMEDIATELY shaking their heads in agreement, as if to say, "finally, someone gets it." Beautiful women repeatedly tell me that they can't fully trust other women and that men are often overly possessive. Being the object of hatred is not the only thing that causes beautiful people distress.

Beautiful women live in a state of free-floating anxiety once they notice that very first wrinkle or other sign that their beauty is fading. Thus, beautiful women live in a state of permanent insecurity, while others project upon them consummate confidence. Think about that state of affairs. On the inside a beautiful woman is insecure, while the entire world projects upon her consummate confidence.

If a woman happens to be smart *and* beautiful, the jealousy and competitive anger she will receive from other women in her chosen profession or school are an ever-present source of hindrance and prejudice. During my hospital internship, I recall a highly competitive female psychologist who treated attractive female interns, doctors and patients with disdain, while trying to seduce every good-looking male doctor and intern on "her" floor. Like almost all sexual harassment, it was this non-descript looking female psychologist's position of power that she leveraged against those people in her midst who did not enjoy the institutional power she used to satisfy her venal desires and pathological defects.

So, while Doctors Dion, Berscheid, Walster, et al. made a persuasive case that "What is beautiful is good," the fact of the matter is, the life experiences of beautiful people can be summed up by this simple statement: "Beautiful people are the people we love...to hate." [11]

THE SOCIAL ENGINEERING OF BEAUTY

There is a famous Supreme Court opinion on obscenity where one of the justices observed this, "I know it when I see it." [12] It is sort of that way when it comes to beauty. Like everything else involving visual image, however, the devil is in the details. And for those of us researchers who dared to ask the question, what is it about certain visual stimuli that encourages people to label them beauty? We entered a world of unbelievable complexity and nuance.

There are a lot of myths about beauty that people assume to be true because they've heard them their entire lives, and they appear to have some face validity. The first so-called truth is that beauty is in the eye of the beholder. That simple declaration is comprised of two parts. The first part is that it is the observer who imbues the judgment of beauty upon any visual stimuli. The second assumption is the motivation behind repeating the myth in the first place, namely, that each observer has his or her own unique and learned standards of beauty. A related notion (actually an ideology) floating around these days is that beauty is purely a social construct; that is, beauty is 100% culturally defined.

If you buy this ideology, then you believe that social engineers can convince men that unattractive women are actually beautiful. According to these ideologues, all it takes to accomplish this is to change our collective social learning. These myths and ideologies spawned by progressive social engineers are utter nonsense because they have no basis in fact.

Beauty is a sociobiological marker that is related to overall health, reproductive viability and youthful vigor. Therefore, we humans are hardwired to find the following component parts of reproductive viability, overall health and youthful vigor to be beautiful; that is, beauty can be defined by a number of

---

[11] To read more about forensic cases involving violence against beauty, please see MURDEROUS RAGE, available here: https://gumroad.com/dranthonynapoleon
[12] Jacobellis v. Ohio, 378 U.S. 184 (1964). Justice Potter Stewart.

objective qualities *unrelated* to social constructs, learning or individual taste. These include: Symmetry, mathematical balance, proportional contours, homogeneity of skin color and texture, and ratios between areas of the human body (In addition to tens of thousands of other definable characteristics that are EACH worthy of Ph.D. dissertations).

Eyes that are too small, mouths that are too big, dramatically big or asymmetrical noses, skin that is mottled with red and brown flaky patches or covered in erythematous bumps, are unattractive no matter where you go on earth when compared to smooth-skinned, proportional and symmetrical faces.

Men who are broader at the shoulders and narrower at the hips with chiseled, symmetrical features are consistently judged to be more attractive than men with narrow shoulders, broad hips and amorphously contoured and mottled faces. The folly promoted by progressive social engineers of socially engineering unattractive into beautiful is like the alchemist's dream of turning lead into gold, or making cruciferous vegetables as tasty as foods loaded with sugar and fat.

I've never seen nor even heard of a single cosmetic or plastic surgery patient that requested her nose be reshaped into an asymmetrical nose, have their smooth skin roughened or their proportional ears modified into large flapping appendices on the sides of her head. I've never seen nor heard of a cosmetic surgery patient ask for her jowls to be purposefully drooped or the skin on her neck to be transformed from smoothly draped skin into a turkey neck.

And the reason I have never experienced nor heard of such makeover requests is NOT because social engineers have failed to educate the masses on the inherent beauty of a turkey neck, droopy jowls, heavily creased facial wrinkles or an asymmetrical nose, but because these visual stimuli are simply unattractive. In large measure they connote an absence of health, reproductive viability and/or youthful vigor, in addition to violating any number of beauty parameters related to contour, symmetry and ratio-based configurations.

The ONLY exceptions to the above referenced patterns regarding the requests of cosmetic surgery patients involve the witness protection program and the occasional, but rare, request by various intelligence agencies to make an agent unrecognizable using plastic surgery. Even then, every effort is made to maintain the agent's current level of attractiveness.

Apologists are quick to note that body piercing and similar modifications, including tattooing, violate universal standards of beauty, and yet are considered to be beautiful. Not so. If you study the practitioners of tattooing and body piercing, you will find these body artists adhere to the same principles of symmetry, proportion, balanced contours, and use of color as do cosmetic surgeons. In other words, the best practitioners of body modification science and art universally recognize unattractive body modifications, piercings and tattoos.

Another apologist's retort to the sociobiological universals of beauty is that standards of beauty change, thus connoting a cultural learning dynamic that helps to define beauty. Yes, *ethereal* standards of **faddish beauty** do change from time to time and culture to culture, but the dispositive features of visual stimuli we scientists refer to as beauty do not.

Take, for example, the actresses of the 1950s, like Marilyn Monroe and Ava Gardner, and compare them to the actresses of the 21st century. While the more recent crop of beauties do have a "different" look, their waist-to-hip ratios are virtually identical to their 1950s counterparts. The same is true regarding the current trend toward larger derrieres; again, waist-to-hip ratios remain set at what is a universal attractiveness coefficient ideal.

Professor Devendra Singh at the University of Texas believes the waist-to-hip ratio may be the most powerful sexual trigger of all, and what strengthens her theory is the fact that this ratio has recently been recognized as a key indicator of health. Read what professor Devendra Singh has written about the subject of waist-to-hip ratios:

> *"The waist is one of the distinguishing human features, such as speech, making tools and a sense of humor. No other primate has one. We developed it as a result of another unique feature – standing upright. We needed bigger buttock muscles for walking on two legs."*
>
> *In America over the past 50 years, ideas of beauty have varied a lot. A few years ago a popular idea was that the ideal shape was going*

to be androgynous. Between 1955 and 1987 the waist-hip ratio of Miss America contestants and Playboy playmates varied only between 0.68 and 0.71. The young women may have been tall and slender, but they maintained female curves, i.e., waist to hip ratios.

Professor Singh found that males had a clear reaction to different waist-hip ratios. In a survey of 106 men aged 18 to 22, the favorite was a female of average weight with the classic hourglass figure. Not only were such women rated as young, sexy and healthy, they were also seen as ideal for childbearing.

The young men regarded the underweight women - defined as women of 5ft 5in weighing less than 90lb - as 'youthful' but not particularly attractive, especially for childbearing. And they viewed the overweight - women of 5ft 5in weighing more than 150lb - as unattractive, but more suitable as prospective mothers. Men of all ages agreed with these findings - thus bearing out her theory of the waist-hip ratio. So perhaps, as well as fashion's dictates on uplift bras, we may one day see the return of the corset. [13]

Norma Jean, aka, Marilyn Monroe (1950s)

---

[13] Singh, Devendra (1 August 1993). "Adaptive significance of female physical attractiveness: Role of waist-to-hip ratio." Journal of Personality and Social Psychology. 65 (2): 293–307. doi:10.1037/0022-3514.65.2.293.

Scarlett Johansson (2010)

Yes, relatively large derrieres are all the rage in early 21st century America. This trend in the popularity of relatively larger derrieres is the result of what researchers, including myself, have termed "the browning of America."

As demographic shifts occur, wherein Sub-Saharan African and Hispanic genotyped women become a larger percentage of the population of the United States and elsewhere, their natural body types will become more prevalent in the media in order to capitalize upon the Black and Hispanic market.

Caucasian and Asian women striving to "get in on" the popularity of more pronounced derrieres are resorting to prosthetics worn under their clothes, exercising their derrieres in hopes of achieving a more full and shapely butt, and going so far as to undergo a plastic surgery procedure known as "The Brazilian Butt Lift." Despite the trend toward relatively larger derrieres, universal standards of beauty, as evidenced by the waist-to-hip ratio, have not changed. To quote singer and song writer Sir Mix-a-Lot:

I like big butts and I cannot lie

You other brothers can't deny

*That when a girl walks in **with an itty-bitty waist** (a specific reference to the waist-to-hip ratio)*

And a round thing in your face...[14]

Sir Mix-a-Lot's *Baby Got Back* song is only one of several songs created by African-originated artists that celebrate the derriere. I dare say that Sir Mix-a-Lot probably knows nothing about scientific studies on waist-to-hip ratios and has never heard of Dr. Singh and sociobiology. But as the old saying goes, "You don't have to be a weatherman to know which way the wind blows," was never made clearer when this artist inserted the lyrics, **"with an itty-bitty waist."** Sir Mix-a-Lot said this in an interview on the subject of butt songs and Sub-Saharan Africans:

> *"There's always butt songs. Hell, I got the idea sitting up here listening to old Parliament records: Motor Booty Affair. Black men like butts. That's the bottom line." The song is part of a tradition of 1970s–90s African-American music celebrating the female posterior, including "Da Butt," "Rump Shaker," and "Shake Your Groove Thing." [15]*

As a rule, large, flabby, asymmetrical and mottled skinned derrieres are judged to be unattractive no matter who you are or your genotypic reference group. But it is not just the derriere that adheres to universal standards of beauty, regardless of size.

For example, white, evenly spaced and aligned teeth are judged to be more attractive than yellow or brown snaggletooth smiles no matter where you go on earth; and no, it is not because social engineers have enforced a white-toothed (rooted to white privilege) hegemony on the part of the cosmetic dentist bourgeoisie, that have oppressed the yellow and brown snaggletooth citizens among us, or that white privilege has bled over into the dental world.

---

[14] Sir Mix-a-Lot (1992). Baby Got Back, from the Album: Mack Daddy.
[15] Aubry, Erin J. (2003). "The butt: its politics, its profanity, its power." In Edut, Ophira. Body outlaws: rewriting the rules of beauty and body image (2nd ed.). Seal Press. p. 30. ISBN 1-58005-108-1.

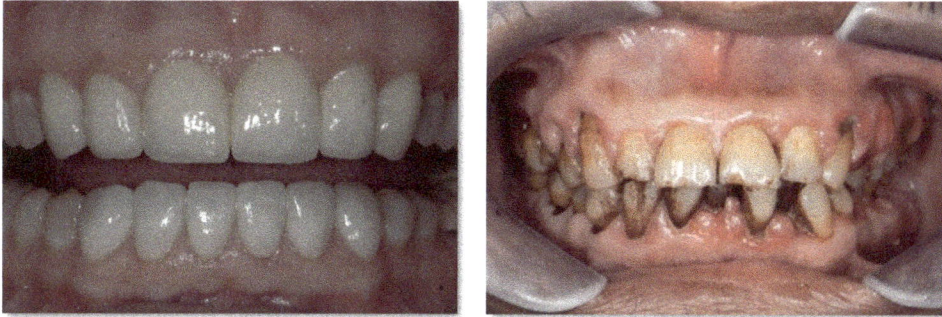

Among some Sub-Saharan African originated people, gold front teeth are all the rage, as are diamond-studded teeth. Such modifications to one's dentition are called "grills." Nevertheless, those grilled teeth are evenly spaced, the gold is shiny, and the most attractive grills have diamonds that are clean (few if any inclusions). *"But if only the media would highlight charming brown, snaggletooth spokespersons like the mouth pictured above on the right, we'd all change our views of what constitutes a beautiful smile and wouldn't be so taken with the photo on the left."* Such tripe is just so much politically correct and ideologically driven balderdash that it borders on insanity. Welcome to the delusional world of the 21st century social engineer ideologue.

The good news is that since we now have a more complete understanding of the wired-in fundamentals of beauty; surgeons, dermatologists, aestheticians and others in the beauty business can improve almost anyone's appearance.

Mottled skin can be treated with lasers that correct broken blood vessels and sunspots, and reduce fine lines and wrinkling; thus, the patient becomes more attractive. Asymmetrical or disproportional noses can be reshaped into symmetrical and proportional features on our face; thus, we become more attractive. Ears that protrude and may be a source of insecurity for our children can be modified; thus, saving our children from a life of self-consciousness. Misshapen body piercing can be replaced with a more beautiful version and body artists can redesign and cover up bad tattoos.

Years ago, I originated the phrase that you've probably seen in plastic and cosmetic surgery advertisements: "Look as good as you feel." When our understanding of beauty is empowered by our medical technology, there is no reason to resign oneself to being unattractive, or to engage in the foolishness of trying to socially redefine beauty so as to make those less attractive among us more socially desirable. Be thankful that we now have medical technology that empowers our understanding of beauty. Today everyone, no matter who you are, can improve your looks.

BEAUTY AS CURRENCY

One of the best predictors of a man's income is the attractiveness of his wife or girlfriend (more so his girlfriend than his wife). As a general rule, as a woman's attractiveness increases, so does the income or net worth of her husband or boyfriend. Within that general rule of what I have termed The Rich Man/Pretty Woman matchup, we see evidence of an interesting feature of beauty. Beauty can be bartered for goods and services without first converting it into dollars. Beauty is also a valuable commodity that can be easily converted into fiat currencies, that can then be spent for goods and services.

Beauty, in and of itself, is a valuable commodity recognized all over the world. However, should one wish to convert the commodity of beauty into dollars, Euros, rubles, etc., then beauty, like every other currency is vulnerable to exchange rates.

For example, the commodity of Ukrainian beauty has more value when it is converted into dollars than when it is spent using the Ukrainian hyrvnia or converted into Russian rubles. Economic theory would predict that because the commodity of beauty can function as a currency, and there are such things as exchange rates, there would be a market for beautiful Ukrainian women in the West because of the favorable exchange rates. This is, in fact, the case.

The Ukrainian Hyrvnia

As of the writing of this book, one American dollar was equal to 26.05 Ukrainian hryvnia.

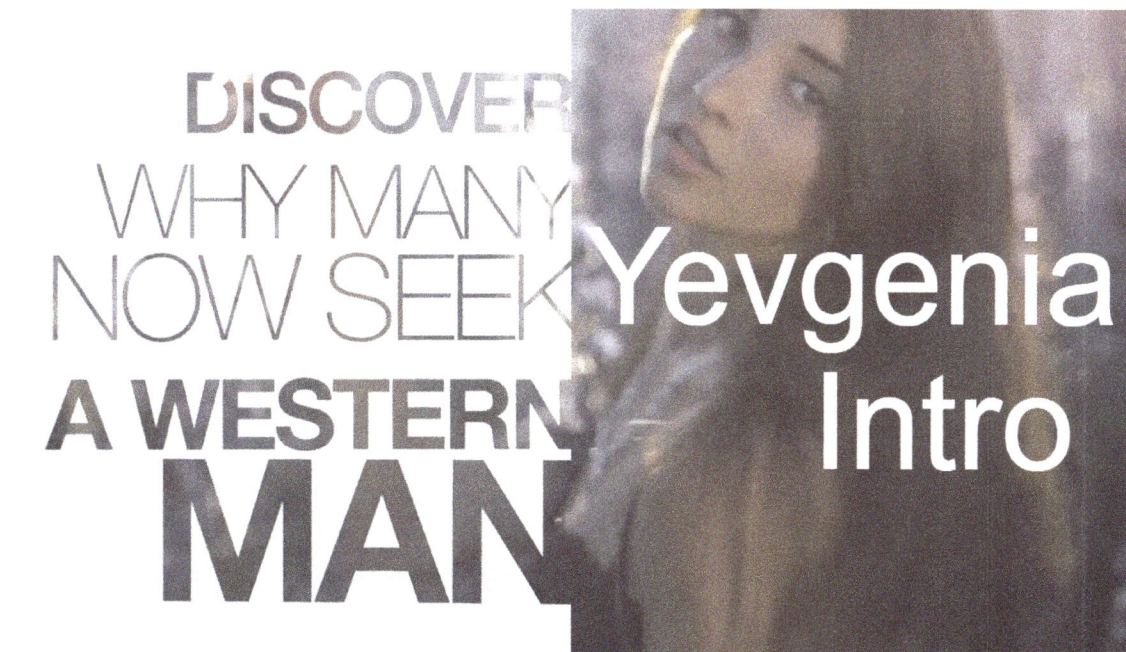

One of the more interesting and powerful dynamics in all of economics is the concept of money supply. In past generations, women who possessed beauty levied tariffs and kept a tight rein on their money supply, i.e., their beauty. In other words, beauty's value was kept at a premium because its supply was limited in terms of how easily a person could buy access to it. World-renowned anthropologist Margaret Mead once remarked, "The courtship of desirable women costs money." This dynamic kept the value of beauty relatively high and its supply relatively low in America and much of the Western world in the past.

Beginning sometime in the 1960s, women who called themselves feminists encouraged all women, including higher valued beautiful women, to lower their tariffs and provide easy access to their currency. In other words, all women, *especially beautiful women*, were encouraged by *unattractive* women to increase THEIR money supply, i.e., "print" more of THEIR money.

Feminists disproportionately affected beautiful women because beauty is a higher valued currency, and beautiful women have more of it when compared to the money supply and value of the currency belonging to relatively unattractive women, i.e., feminists. As with all commodities, supply and demand set the price of the commodity in question.

Of course, the end result of "easy money," whether we are talking about beauty or dollars, is a devalued currency that has much less buying power than it once had. When compared to 1970, for example, $20 USD is now (2017) worth approximately $3.48 USD. The devaluation of the currency of beauty would predict that beautiful women would find it exponentially more difficult to parlay their beauty into financial security. In other words, $20 no longer buys what it once could buy. The data bear this prediction out.

A 2013 PEW Research study found that 40 percent of American women are the primary or sole breadwinners. Seventy-one percent of men make less money than do their wives as of 2013. In 1960, only 4 percent of women made more money than their husbands.

Now to be sure, the broad-spectrum cultural revolution of the 1960s impacted gender roles and employment in ways unrelated to beauty that most certainly account for some of the PEW Research findings. Nevertheless, it is the devaluation of the currency of beauty that has accounted for the trend toward women buying their own engagement rings. Jewelers report that with each passing year, fewer and fewer men are buying engagement rings for women. Part of that, of course, is that men no longer commit within the confines of marriage. In as much that commitment, as evinced by the engagement ring, was a tariff levied on access to beauty; as tariffs were dropped, so was commitment and the cost of courtship.

As the currency of beauty becomes less valuable with each passing year, one would predict that more and more of that currency would go into circulation. In fact, that is exactly what has happened. Beaches in America are strewn with women wearing thong bikini bottoms, see-through tops and bikini bottoms cut so low that in order to wear them, women choose to wax their vulva.

It is not just swimwear, of course. It is quite common to see female bicyclists wearing skirts and dresses, flashing their underwear with each peddling cycle. Once upon a time, when the money supply was kept under control, going without a bra was scandalous; today it goes unnoticed, as do exposed panties flashing beneath short skirts. It is ironic, but predictable, that even as the currency of beauty floods the market, men no longer value the glut of cheap dollars in the form of sex that can be bought with no money down and a zero percent interest rate.

Feminists turned economic self-interest on its head when they persuaded beautiful women to devalue their currency by lowering tariffs and printing more money. What feminists did to beautiful women would be analogous to a situation where the lowest valued currency nation persuaded the United States to devalue its currency, so that the U.S. dollar reached near parity with their currency.

BEAUTIFUL THINGS

Beauty's allure is not limited to beautiful people. The diamond industry is a good example of mankind's love of inanimate beauty. The diamond industry employs 10 million people worldwide. Backing out the industrial use of diamonds, we are left with a beauty-related jewelry business worth in excess of $72 billion. Not only do people love diamonds, rubies, emeralds and sapphires, but precious metals such as platinum, gold and silver are attractive to human beings.

It is fascinating to study the things jewelers choose to model their jewelry after. One of the most prolific models used in the jewelry industry is the ladybug. Adorned with her curvilinear body and covered in dark spots on a red background, this little insect has been the artistic inspiration for cups, saucers, fabrics, pins, broaches, rings and necklaces.

The Ladybug, aka "Aphid Dragon" Earrings and Pendant Set

The humble ladybug provides us a marvelous case study in visual beauty versus substance. The ladybug goes by another name you may not have heard of: "The Aphid Dragon." Yes, this beautiful little model is also a vicious little killer. The ladybug captures plant-eating aphids and, while their victim is still alive, this stark raving beauty—a model for all sorts of human-adored artifacts—vacuums out her victim's bodily juices until they die a slow and miserable death. Now, I ask the reader to compare the ladybug's behavior to that of the praying mantis. This insect identifies its prey with its bulging eyes, which are attached to part of its tiny little head, as it lunges out with its long, spindly arms and captures its prey. Which insect is more disturbing to you?

The Praying Mantis, aka "Miss Ugly"

A hint as to which insect is more disturbing can be found in this question: When is the last time you saw a gold broach in the form of a praying mantis or a child's bib with a praying mantis emblazoned upon it?

Consider the beautiful butterfly's popularity as a jewelry model:

Compare the beautiful butterfly's popularity to the black widow spider's popularity as a jewelry model:

Black Widow Spider Doing What She Does Best

As you can see, the black widow spider is a popular model, especially at Halloween when the goal is to creep people out, or when the black widow ring on your finger or tattoo below your panty line is meant to convey your darker side. Try handing out a piece of candy in the shape of a bear (Gummy Bears™) or a piece of candy in the form of a realistic-looking spider. Study the response you receive. Nature provides us many examples of the discordance between visual images and behavior. For example, humans tend to think bears are cuddly and adorable because they have large, round, furry heads.

A Pair of Cute Brown Bear Cubs

Unlike cute bear cubs, human beings are not that fond of baby rats because of their pointy noses, big bulging eyes and hairless pink skin; nor are most people into piglets because their heads are approximately the same size and shape as their body.

Never mind that bears view you as dinner or that pigs are as intelligent as dogs, or certainly more intelligent than horses. None of that matters, because rats, pigs, and most creepy crawly things have a beauty deficit. Children all over America go to sleep with their teddy bears, and everyone loves panda bears. Pandas are one of the most popular animals on earth (they aren't even bears). Horses are majestic, vibrant and beautiful, but hippos are—well, they're hippos.

Beauty and the Beast

The sellers of this world know human beings love beauty in whatever forms it may take. Whenever it is possible, things are made to be more beautiful, because beauty sells. Nowhere can this be seen more clearly than at your local grocery store. Stroll down the produce aisle and you will see waxed cucumbers, dyed oranges, polished apples and fluffed-up greens. Botanists tinker with the genetics of fruits and vegetables so that they will look more beautiful, NOT necessarily taste better.

If you walk down the cereal aisle, you'll see grains for sale in beautiful boxes that often cost more to manufacture than the product inside the box. This is because the design of the package—that is, its beauty—is the determinative factor in catching the shopper's eye. The grains inside the beautiful box have a beauty deficit.

Design engineers often override the prescriptions of mechanical engineers. All of us, at one time or another, have purchased an attractive product only to find out later that its functionality was lacking. For years, Volvo manufactured relatively unattractive but very safe cars, while Detroit manufactured beautiful death traps. In 1999, that changed when Volvo went Hollywood and hired design engineers who gave the mechanical engineers in Sweden a run for their money. The S80 was advertised by a television ad that included a voiceover artist repeating the phrase, "Look at my body." The first model years of the Volvo S80 were plagued with transmission problems. But it looked beautiful on the showroom floor.

Finish carpenters are a builder's most valuable assets, because these are the people who create the façade that sells the home. If you are a contractor who is going to cut corners, you'll cut corners on plumbing and electrical items before you will scrimp on finish carpenters. Buyers can't see the poorly designed pipes or the cheap wiring running through their beautiful new homes' façade. But the crown molding? Now that looks beautiful, and buyers can see it.

You might be getting the idea that beautiful things can get you into as much trouble as a beautiful face. If you are thinking that, you're absolutely correct. Whether it's people or things, beauty encourages us to expend our time, effort and money when otherwise we should walk away. Sellers know this. Perhaps this is why the Greeks cautioned: Caveat Emptor.

THE EYES HAVE IT

The eyes are reputed to be the window into the soul. This makes sense in that the eyes actually develop from the neural tube, to wit:

> *"The major development of the eye takes place between week 3 and week 10 and involves ectoderm, neural crest cells, and mesenchyme. The neural tube ectoderm gives rise to the retina, the iris and ciliary body epithelia, the optic nerve, the smooth muscles of the iris, and some of the vitreous humor. Surface ectoderm gives rise to the lens, the conjunctival and corneal epithelia, the eyelids, and the lacrimal apparatus."* [16] [17]

Eyes are also that part of our face that others often look at first, while our eyes are busy returning the favor. Ironic, isn't it? Is it any wonder that the makeup industry, plastic surgeons, eye doctors and security agencies have a keen interest in human eyes?

Our lexicon reveals mankind's fascination with eyes. Consider the following descriptive words and terms that attempt to capture the meaning communicated by the human eye: *Bedroom, Dagger, Evil, Dead, Mischievous, Shifty, Kind, Teary, Steely,* and the list goes on and on.

Eyes and romance are so closely related that man's artistic and cultural expressions are replete with references to eyes. Here are just a few song titles that illustrate this cultural pattern: *Brown Eyed Girl, Betty Davis Eyes, Private Eyes, For Your Eyes Only, Don't It Make My Brown Eyes Blue, Angel Eyes, These Eyes are Crying, Behind These Hazel Eyes, Eye of the Tiger, Behind Those Eyes, Blue Eyes Crying in the Rain*, and many, many others.

---

[16] Ort, Victoria, Ph.D and Howard, David M.D. (2017). *The Development of the Eye.* New York University School of Medicine.

[17] Neural Tube: (in an embryo) a hollow structure from which the brain and spinal cord form. Defects in its development can result in congenital abnormalities such as spina bifida.

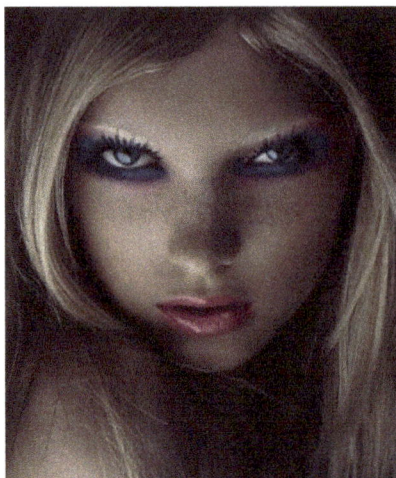

Bedroom Eyes

During the Middle Ages, women used the juice of a berry from the nightshade plant to dilate their pupils. Even in the Middle Ages women understood that when eyes are dilated, their owner becomes more beautiful to men. In fact, the word used to name the chemical derived from the nightshade plant is "belladonna," or Italian for *beautiful woman.*

It is, after all, our pupils' dilation in response to moonlight that makes a moonlit night romantic. As Milan wrote:

> *"I'll be your friend in daylight. I'll treat you as a comrade in every gas-lit*
> *ballroom. But alone, under moonlight, I'll not pretend that I want you*
> *for anything but mine."* [18]

For those of us who research such matters, we have long wondered, what is it about dilated pupils that increase their owner's beauty? The answer is less romantic and more of a lesson in sociobiology.

The basis for the attraction of dilated pupils is that humans' autonomic nervous systems dilate our pupils in response to gazing upon something of interest, particularly sexual interest. Dilated pupils, therefore, serve as a subliminal communication to others that we have a sexual interest in them.

Humans are naturally attracted to those people who make us feel sexually attractive. Thus, dilated eyes are attractive, whether made so by belladonna,

---

[18] Milan, Courtney (2011). *Unlocked (A Turner Series).* Courtney Milan Publisher.

moonlight, candlelight, or because of sexual interest. What this means is that if you really want to gauge another person's sexual interest in you, don't do so over a candlelit dinner where everyone's eyes are naturally dilated; instead, take your date to lunch on a sunny day.

Eyeliner, eyeshadow, mascara, eyelash extensions, and biologicals such as Latisse are designed to grow longer eyelashes or create illusions that make the human eye appear to be larger, more defined, more colorful; in other words, more beautiful. Eye drops that whiten the sclera (whites of the eyes) by constricting blood vessels make the eye look refreshed and younger, along with enhancing the contrast between the iris and the surrounding tissue.

Plastic surgeons perform upper blepharoplasties on patients who have excess skin hooding on the upper eyelid, usually the result of aging and its concurrent loss of skin elasticity. Lower blephs are often designed to remove bags or puffiness under the eyes. An eye with excess hooding tends to look older and less vibrant. Bags under the eyes tend to connote less than optimal health and the loss of youthful vigor.

Colored contacts can change or enhance the natural color of the eye. One fascinating trend made popular in Asia is the circle lens. Here is how one seller of circle lenses named PINKY PARADISE described their product:

> "Circle lens is a cosmetic contact lens, also generally known as big eye circle contacts, which has a wide black ring on the outside that widens your eyes' iris, creating an adorable appearance that you'll fall in love with. Circle lenses were made popular by Ulzzangs ("Best-Face" celebs) like K-pop stars Girl Generation, Hyuna, Krystal and Suzy. Ulzzang is also shorthand for pretty, and you'll look the part in these lovely circle lenses. In Japan, you will find circle lens appearance in many magazines and they are always considered as a "must factor" across all models like Tsubasa and Super star Ayumi Hamasaki. These stars and models are

famous for wearing bright circle lenses to make sure that their eyes sparkle in music videos." [19]

Model Before and After Placing Circle Lenses (plus hair and makeup work)

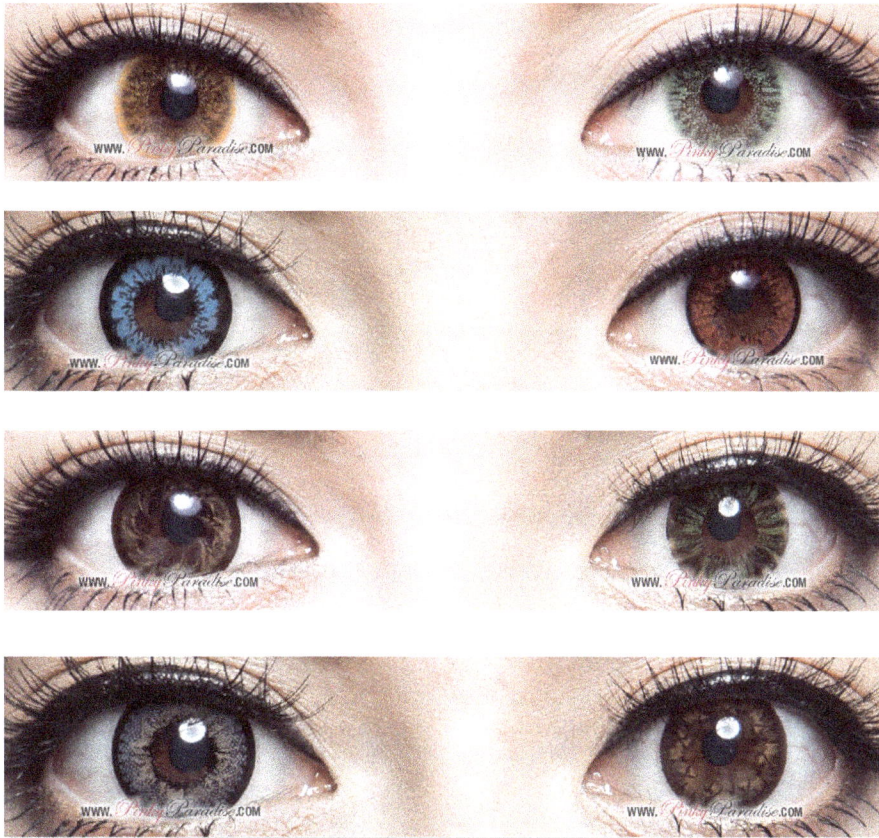

Examples of Circle Lenses

It is noteworthy that the development of the circle lens occurred in Asia, where the people who live there naturally have eyes adorned with an epicanthic fold. This skin fold gives the "Asian eye" its characteristic look.

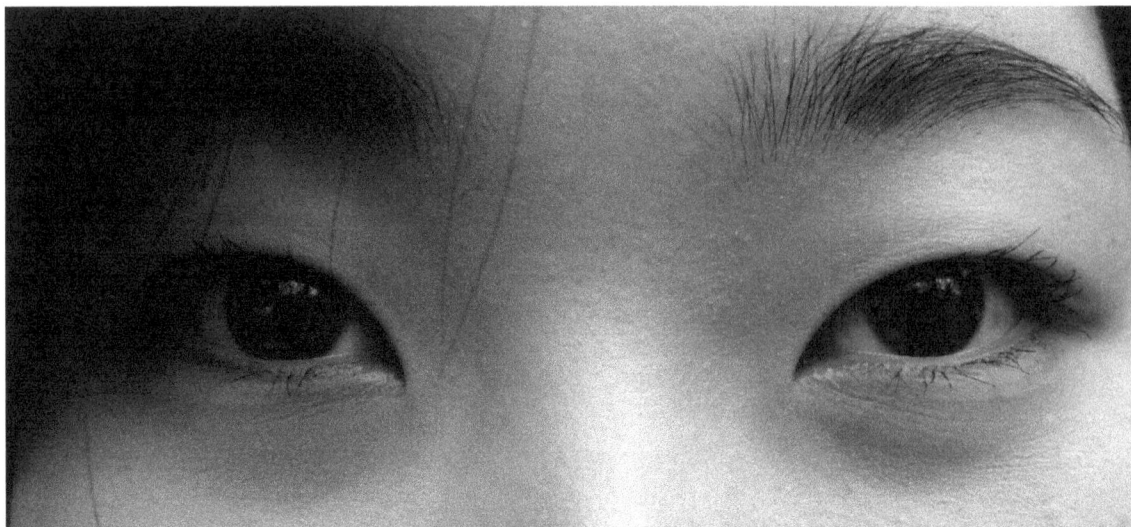

Example of The Epicanthic Fold, Also Known as The Palpebronasal Fold

A few exceptional actors are masters at consciously changing the configurations of the muscles around their eyes in order to convey their characters' emotional and cognitive state of mind. The great actor Sir Michael Caine has said that one of the first tricks he learned, as a young actor, was to not blink during an important scene. By not blinking, Caine said that he could capture the audience's undivided attention.

Caine had discovered that when humans experience strong emotions, our eyes remain open for longer periods of time between blinks. Few of us consciously recognize this, but that doesn't matter because such subtleties in eye behavior have their impact on our unconscious level of awareness, where in many ways that effect is made more powerful because it is subliminal.

Of course, there is the wink that is universally understood to be a private communication. A wink between a man and a woman is most often a flirtation; whereas, a wink between two people in a group of three or more typically conveys to the person winked at that the communication being made to a third person is disingenuous.

Clint Eastwood's character Dirty Harry became famous, in part, for displaying a squinting set of eyes fixated at a criminal as Harry delivered one of his famous movie script lines, e.g., "You got to ask yourself, do you feel lucky, well, do you punk?"

Clint Eastwood, aka "Dirty Harry"

Dirty Harry would not have had such charismatic stage presence, nor would the screenwriter's words have become so famous, had Officer Callahan been blinking rapidly with wide-open eyes as opposed to the steely, squinting stare of Clint Eastwood as he spoke those now-famous script lines.

Marilyn Monroe had bedroom eyes, the result of her eye's natural structural configuration and Marilyn's natural tendency to keep her eyes slightly closed as though she were intently interested in you. Marilyn also suffered from a minor amblyopia, which affected her left eye (see the photo below). This tended to give Marilyn a mysterious and come-hither look. These particular eye configurations, taken in toto, helped to make Marilyn appear to be reactive, alive and sexually vibrant.

Marilyn Monroe

Many people will attest to the fact that they have a sixth sense about being stared at by another person. But if an intermediary piece of technology is sandwiched between the eyes doing the looking and the person being stared at, that sixth sense disappears. In as much that cameras capture our every move these days, it is unnerving to think that behind each camera is at least one set of eyes. When actual prying eyes are directed at us, we act differently. Studies show that by merely placing a painting of a set of eyes on a lunchroom's wall, the people who congregate there become more subdued and more circumspect.

You'll hear some people scoff at those who worry about all those government eyes watching you through all those digital devices. These people say things like, "I have nothing to hide." There is a reason that bathroom doors do not have peepholes and windows have curtains. And just think, in the not too distant future you can rest assured that no matter where you go, even in your own home, someone's eyes will be looking at you while listening to every sound you make.

MIRROR, MIRROR ON THE WALL

Imagine what it was like for early man the first time he gazed upon himself in a still body of water. Still water gave way to mirrors, and mirrors gave way to portraits, then film cameras. Film cameras evolved into digital cameras that are now incorporated into virtually every smart phone. These same smart phones are our ticket to social media's super highway.

Social media promises virtually instant connectivity with any other person on the same digital super highway. The traffic on social media is comprised of words and images. And of all the images traversing the roads of social media, none are so popular as our own images we take of ourselves. Like all things or behaviors that become popularized, we have a name for a self-taken photo: the "selfie."

Taking a Selfie

Now before I explore the psychology of the selfie, I'm going to digress for a moment and talk about people who are so obsessively critical of their appearance, they develop a condition known as Body Dysmorphic Disorder (BDD).

BDD is characterized by an obsessive focus upon a real or perceived body defect or set of defects that unpleasantly intrude into the consciousness of the patient at will, and encourage a compulsion to do things that are designed to compensate for the real or perceived defects.

Forty-two percent of people with BDD spend from 1-3 hours per day looking in a mirror or obsessing over their self-image. Thirty-nine percent of those suffering from BDD spend 3-8 hours per day ruminating over their visual image, while 10 percent of BDD patients spend 8 or more hours per day engaged in a troubled, obsessive and compulsive fixation upon their visual image. With that brief background of BDD in mind, let's get back to the selfie.

I've always thought of any mental condition or diagnosis as best understood as falling on a continuum that runs from traits or tendencies to full-blown psychopathology. Think of this concept this way: obesity falls along a continuum from "I need to lose a few pounds" to being morbidly obese. If BDD is on the extreme end of the continuum, then what is on the other end? And if the average person on the traits or tendencies end of the continuum looks at herself a dozen times a day in a mirror, will the popularity of the selfie move that person toward the BDD end of the continuum?

Plastic surgeons report that the increasing popularity of the selfie has resulted in an increase in the number of people seeking consultations for surgery. Plastic surgeons report that more and more prospective patients bring a collection of selfies to their first consultation in order to point out defects in their physical appearance.

The practice of patients bringing images of themselves or others into their first plastic surgery consultation is well known to plastic surgeons. One interesting question is this: why do so many people not like the way that they look in their selfie? I mean, after all, certainly these same people have seen themselves many times in the mirror, right? And their mirror image didn't motivate them to seek out plastic surgery.

During my psychology fellowship in a plastic surgery practice, I developed a reverse mirror technique in order to prevent post-operative dissatisfaction among patients who tended to be obsessively critical of their looks. The reason I developed this pre-op assist had to do with the reason many people do not like the way they look in a selfie.

When we look in a mirror, we see a reverse image from the one we see when we look at ourselves in a selfie or what other people see when they look at us. To understand this visual phenomenon, look at your left and right hands. We like to think that our hands are the same, like the selfie and your image in a mirror; but if you try to superimpose one hand upon the other, you'll see that they are not the same. Your image as seen in a mirror and the one you see in a selfie are what we call "optical isomers" of one another. Now follow me here, because this is where it gets tricky. By the way, I told you in the beginning of this book that the subject of visual stimuli was a lot more nuanced and meaningful than you ever imagined.

Hands are Optical Isomers of One Another

When you look in a mirror, over time, your brain tends to balance out the asymmetries of your face. It is like your brain's version of your hand lifting the corner of an off-center picture frame hanging on the wall. This balancing act on the part of the brain is why unattractive people seem to become more attractive over time, including our own mirror image. This time phenomenon works because symmetry, or the lack thereof, is a big part of beauty.

Your Brain Tends to Make the Picture on the Left into The Picture on the Right Over Time

But here is where our dissatisfaction with selfies originates. When you look at yourself in a selfie, your brain's compensatory mechanisms are off-kilter, because the vast majority of the time your brain's balancing mechanisms have been conditioned to correct asymmetries in mirror images. Unfortunately, the same balancing mechanisms that improve your mirror image tend to exaggerate asymmetries and other defects in your selfie. Let's refer back to our off-kilter picture frame analogy.

Looking at that off-kilter picture frame's reflection over time means that your brain will even out. Let's say, the right lower corner needs to be raised in order to make the picture frame more balanced. Now imagine looking at a photo of the picture frame. The side off-kilter, i.e., the right lower corner, is now the left lower corner. Therefore, when your brain does its balancing act and raises the right lower corner, it actually makes the picture frame look worse! This same phenomenon is why your selfie looks like what other people see all the time, but looks like a more unattractive version of you. This is why others genuinely think you look better in photos than you do; it is not just that your friends are being tactful.

Try this experiment at home: stand side by side with a close friend in front of a wide mirror. Now look at one another's reflected facial images, and you will see asymmetries and defects in one another that you never noticed before. Now, turn your gaze away from the mirror and look at the other person directly, then turn your gaze back to the reflected image. The differences you will notice are why we tend to be more critical of ourselves when we look at our selfies, so much so that more and more of us seek out plastic surgery to make us feel better about our looks.

The other factor that distinguishes the selfie from reflected images is that selfies are typically published on social media. It is the act of dissemination of one's selfie that exacerbates our critical evaluation of ourselves. I might add that a surprising number of movie stars refuse to view themselves in their own movies because of the dynamics I've just expounded upon for my readers. Historically, gangster character actor George Raft refused to watch even a brief clip of one of his many famous gangster movies of the 40s and 50s.

Raft was a frequent guest on the classic Tonight Show with Johnny Carson. Carson was fond of showing his audience clips of Raft's more memorable roles as a villain. As the clip played on the in-studio monitors, Raft turned his head away and shielded his eyes with his hand, as if he were a vampire turning away from the sun. Many of today's most famous actors share Raft's angst over viewing movie selfies; these include Johnny Depp, Angelina Jolie, Julianne Moore, Megan Fox, and many, many others.

Gazing at one's image, and the psychological problems it causes, have been written about for hundreds of years. Perhaps no other story illustrates these problems better than the myth of Narcissus. If you are wondering, yes, this is where Sigmund Freud got the name for Narcissistic Personality Disorder. And no, the personality disorder of narcissism is not merely a synonym for being self-centered, egoistic, or any of the other misuses of the term "narcissism" made by lay people.

According to Greek mythology, Narcissus was walking by a lake one day when he decided to drink some water. As he knelt down to drink, he noticed his reflection in the water and immediately became smitten by the beauty he saw. He was so mesmerized by his own image that he became entranced by the reflection of himself, to the exclusion of anyone or anything else.

Narcissus by the Pool

Narcissus' self-love became a source of misery for him, because at some point he realized that he could not obtain the object of his desire. He became so forlorn that he died at the banks of the same lake where he first saw, and then fell in love with, himself. The myth continued into the afterlife where, according to the story, Narcissus can still be found to this day admiring his reflection in one of the rivers of hell named Styx. It would be a river in hell now, wouldn't it?

BEAUTY'S RELATIONSHIP WITH BEAUTY

One of the more fascinating aspects of the beauty phenomenon is that beautiful people often choose significant others who are less attractive. The confound that must be accounted for in this pattern of behavior is that beautiful people are more likely to pair off with people less attractive simply because of the statistical low probability of finding someone as equally beautiful. Nevertheless, even when we account for this particular statistical confound, beautiful people do not appear to place as much value on beauty when selecting a significant other as do average or unattractive people.

Elisabetta Gregoraci and Flavio Briatore

The disparity in attractiveness often seen between beautiful people and their mates creates a number of interesting dynamics. First, there is a gender difference. Beautiful women are just as likely to choose less attractive men to be their boyfriends as their husband; whereas attractive men tend to date beautiful women, but are more likely to marry someone not quite as attractive as his girlfriends.

Another gender difference involves the beautiful woman who is paired with the significantly less attractive man. In those circumstances, the unattractive men in these beauty-paired relationships often become jealous, sometimes pathologically so. The pattern is not so dramatic when the genders are reversed.

The probability of beautiful women becoming victims of their unattractive boyfriends is higher than in the population as a whole. The underlying dynamic that accounts for this disparity is that the unattractive male, absent some compensatory quality like special brilliance or talent, and specifically ruling out wealth, tends to feel habitually insecure in these beauty-paired relationships. That insecurity manifests as possessiveness and resentment over the fact that their beautiful significant other is continually "hit" upon by other men and is continually solicited for her beauty, which means she is continually receiving threatening attention. Resentment, by its very nature, works like compound interest and continues to build until the breaking point. Attractive men and their relatively unattractive wives, on the other hand, often create long-lived and happy marriages.

Another interesting beauty dynamic involves beautiful women and animal welfare. While average or unattractive women tend to be attracted to purebred cats and dogs (I have never seen a beautiful animal breeder), beautiful women tend to focus upon orphaned and/or non-descript animals. From actresses of the classic era like Doris Day and Brigitte Bardot, to more recent actresses like Pam Anderson and models like the Barbie Twins, beautiful women appear to be the orphan and non-descript or unattractive animals' best friends. Our choice of companion animals can tell us even more valuable information about beauty and human behavior.

Brigitte Bardot with Rescued Dog

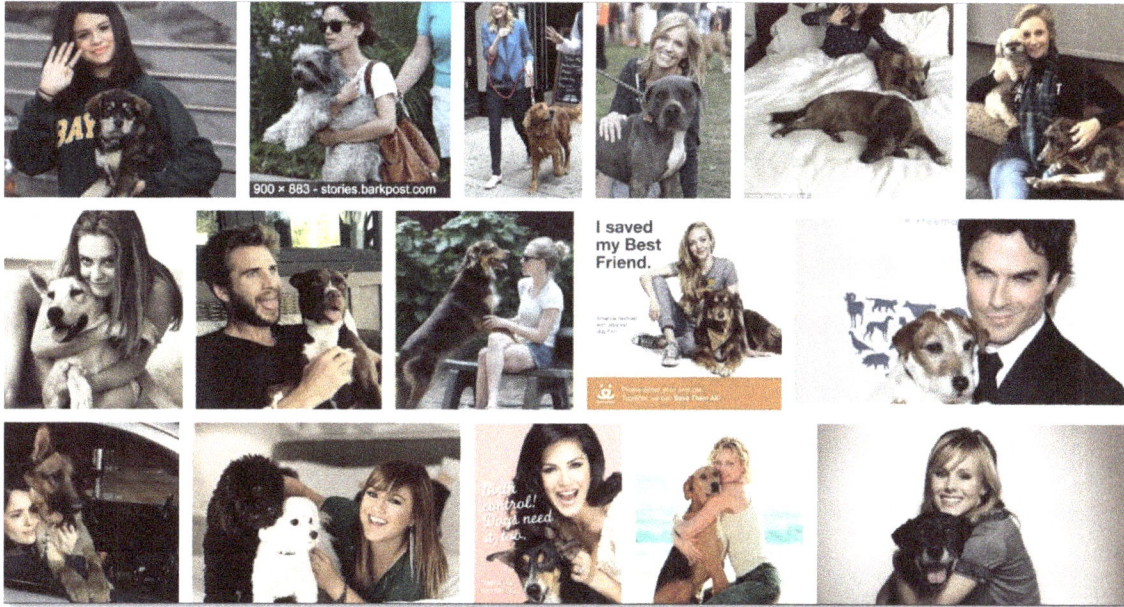

Various Actors and Actresses with Rescued Animals

Westminster Kennel Show (Photo Credit: SUSAN WATTS/NEW YORK DAILY NEWS)

Average or unattractive people tend to choose companion animals that look like what they want to look like; whereas beautiful people tend to choose companion animals, as we have already stated, based upon the vulnerability of the animals in question, not their looks. This means that relatively unattractive or average people tend to choose companion animals based upon how that animal enhances their ego and/or the animal that possesses a "special" (I'm special, too) look. The exception to this rule involves highly compassionate and/or empathetic people, regardless of their level of attractiveness.

Yes, many people tend to look like their companion animals, unless we are talking about beautiful people. Exceptions to this pattern of behavior include libertine trust fund paparazzi women who choose animals as accessories, not unlike how these people would choose a fashionable handbag to complement their wardrobe.

Having lived in Manhattan for a time, I reflect upon a popular saying related to beauty that goes like this, "The secretaries dress like models and the models dress like secretaries." Models, and other beautiful women who work in beauty-intensive businesses, are known to dress down and avoid makeup when off stage; whereas, average-looking women tend to enhance their looks with makeup and beauty enhancing clothes even when running errands.

The dynamic at work in all of these examples is that beauty has less conscious value for beautiful people than it does for average or unattractive people. This suggests that beautiful people, especially women, don't really value that one quality that, in many instances, defines them.

Perhaps this devaluation of beauty itself is why beautiful women are notoriously insecure and lack as much genuine confidence as one might expect. It also may explain, at least in part, why beautiful people have less of a need to possess beauty in other people or things, because for one thing, they already have beauty, and for another, they do not place as much value on beauty as an average or unattractive person. Perhaps this is because beautiful people know the truth about beauty, a truth that only someone with beauty can understand.

And to make matters even more ironic and nuanced, the fact that beautiful people don't necessarily value beauty enrages others who crave what the seemingly unappreciative beautiful people take for granted. Once again, we see confirmation that beautiful people are the people we love to hate.

PLASTIC SURGERY

Elective aesthetic surgery is the only surgical procedure that is solely motivated by psychological factors. Elective aesthetic surgery is to be distinguished from reconstructive surgery, which is purposed to correct an anomaly or defect, e.g., a scar, upper eyelid skin that is disrupting one's vision, excessively large breasts that are causing the patient back pain, or varicose veins that are causing circulatory problems. Now of course, reconstructive and aesthetic surgeries are not mutually exclusive and are often complementary to one another. For example, an upper blepharoplasty that helps one to see better will also make the patient look better.

I am now going to provide data that reflects American's utilization of plastic and cosmetic surgical and dermatological services. I will cover 2013 data, and then show you 2016 data in order to give the reader an understanding of trends. Plastic and cosmetic surgery choices and trends provide a very good sociobiological snapshot of the zeitgeist of Americans and their culture. Most of what you will learn in this section applies to not only America, but also much of the Western world.

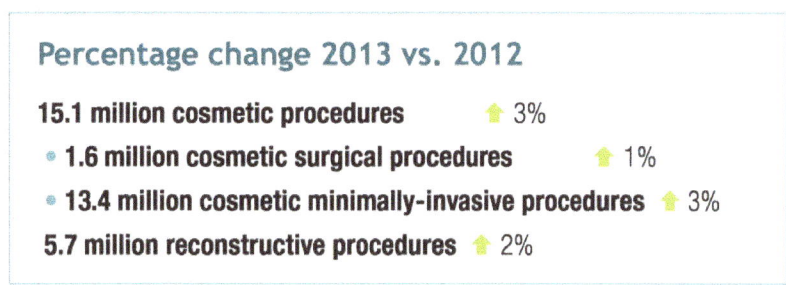

GRAPHIC 1

Percentage change 2013 vs. 2012

15.1 million cosmetic procedures   ↑ 3%

• **1.6 million cosmetic surgical procedures**   ↑ 1%

• **13.4 million cosmetic minimally-invasive procedures**   ↑ 3%

5.7 million reconstructive procedures ↑ 2%

Percentage change 2016 vs. 2015

17.1 million cosmetic procedures ⬆ 3%

· **1.7 million cosmetic surgical procedures** ⬆ 4%

· **15.4 million cosmetic minimally-invasive procedures** ⬆ 3%

5.8 million reconstructive procedures ⬆ no change

GRAPHIC 2

2013 Top 5 Cosmetic Surgical Procedures

	2013 vs. 2012
Breast augmentation (290,000)	⬆ 1%
Nose reshaping (221,000)	⬇ 9%
Eyelid surgery (216,000)	⬆ 6%
Liposuction (200,000)	⬇ 1%
Facelift (133,000)	⬆ 6%

GRAPHIC 3

2016 Top 5 Cosmetic Surgical Procedures

	2016 vs. 2015
Breast augmentation (290,000)	⬆ 4%
Liposuction (235,000)	⬆ 6%
Nose Reshaping (223,000)	⬆ 2%
Eyelid surgery (209,000)	⬆ 2%
Face Lift (131,000)	⬆ 4%

GRAPHIC 4

2013 Top 5 Cosmetic Minimally-Invasive Procedures

	2013 vs. 2012
Botulinum toxin type A (6.3 million)	⬆ 3%
Soft tissue fillers (2.2 million)	⬆ 13%
Chemical peel (1.2 million)	⬆ 3%
Laser hair removal (1.1 million)	⬇ 4%
Microdermabrasion (974,000)	no change

GRAPHIC 5

2016 Top 5 Cosmetic Minimally-Invasive Procedures

	2016 vs. 2015
Botulinum toxin type A (7.0 million)	⬆ 4%
Soft tissue fillers (2.6 million)	⬆ 2%
Chemical peel (1.3 million)	⬆ 4%
Laser hair removal (1.1 million)	⬇ 1%
Microdermabrasion (775,000)	⬇ 3%

GRAPHIC 6

2013 Top 5 Reconstructive Procedures

	2013 vs. 2012
Tumor removal (4.4 million)	⬆ 5%
Laceration repair (254,000)	⬇ 13%
Maxillofacial surgery (199,000)	⬇ 5%
Scar revision (177,000)	⬆ 4%
Hand surgery (131,000)	⬆ 6%

GRAPHIC 7

2016 Top 5 Reconstructive Procedures

	2016 vs. 2015
Tumor removal (4.5 million)	no change
Laceration repair (253,000)	no change
Maxillofacial surgery (202,000)	⬆ 1%
Scar revision (181,000)	⬆ 1%
Hand surgery (135,000)	⬆ 4%

GRAPHIC 8

Breast augmentation continues to be the top cosmetic surgical procedure and has been since 2006. Silicone implants were used in 72%, and saline implants in 28%, of all breast augmentations in 2013.

Cosmetic surgical procedures, not among the Top 5, with notable gains in 2013 include:
- Male breast reduction (gynecomastia) – up 11%
- Tummy Tuck – up 5%
- Neck lift – up 6%

GRAPHIC 9

Breast augmentation continues to be the top cosmetic surgical procedure and has been since 2006. Silicone implants were used in 84%, and saline implants in 16%, of all breast augmentations in 2016.

Cosmetic surgical procedures, not among the Top 5, with notable gains in 2016 include:
- Buttock augmentation w/ fat grafting – up 26%
- Lower body lift – up 34%
- Labiaplasty – up 39%

GRAPHIC 10

Facial rejuvenation procedures experienced the most growth, as 2013 marked the highest number of botulinum toxin type A injections to date, with 6.3 million injections!

GRAPHIC 11

Facial rejuvenation procedures continue to experience growth, as 2016 marked the highest number of botulinum toxin type A injections to date, with over 7 million injections!

GRAPHIC 12

Tumor removal is overwhelming the top reconstructive procedure performed every year with more than 4.4 million procedures in 2013!

GRAPHIC 13

Tumor removal is overwhelming the top reconstructive procedure performed every year with more than 4.5 million procedures in 2016!

GRAPHIC 14

I want the reader to take note of the fact that the overall number of plastic and cosmetic surgery procedures continues its upward trend over the past decade, with an increase of approximately two million procedures from 2013 through 2016. The media almost exclusively focus upon aesthetic procedures such as breast augmentation, but what these this data demonstrates is that the "reconstructive" part of plastic surgery, by far the most popular, involves purely medical issues, e.g., tumor removal. I mention that because peculiarly, the medical part of plastic **surgery** is overlooked. I noticed this peculiar form of denial of the surgical part of plastic surgery during my fellowship. Some patients, for example, expressed curiosity as to why we would like to obtain an EKG or Echocardiogram before performing surgery, after all, it was *just plastic surgery.* Take a look at graphics 3 and 4. Notice that for all of purported emphasis upon androgyny and the promotion of average or, frankly, unattractive socially engineered role models, the number-one elective aesthetic surgical procedure remains breast augmentation.

A similar disconnect between what social engineers push as the desired aesthetic norm and what reality tells us is important, is illustrated in graphics 4 and 5. Wrinkles, a seemingly inevitable consequence of aging, and from progressive social engineers' perspectives a "good" thing, are being erased using plastic and cosmetic surgery in addition to any number of light, filler or chemical-ablative related technologies. Botulinum toxin, e.g. Botox™, continues to be very popular, along with soft tissue fillers such as Juvederm™, Restylane™ and Radiesse™, just to name a few. The Obagi Blue Peel™ (a modified trichloroacetic-acid peel) continues to be popular.

The trend toward cosmetic modification of looks has only become more popular since images of our faces and bodies became our social media calling card. Photo-retouching is the old term, Photoshopping is the "new" term. However, as the following data illustrate, more and more people want to modify their "objective image."

- In 2016, **Americans spent more than 15 billion dollars** on combined surgical and nonsurgical procedures for the **first time ever.**

  $15⁺ Billion

  o There was a **1.5 billion dollar increase** in expenditures over the past year alone.

  o Surgical procedures accounted for 56% of the total expenditures in 2016 and nonsurgical procedures accounted for 44%.

  56%

- Surgical procedures were up 3.5% in 2016.

- The surgical procedures that saw the most significant increases in 2016 include:
  Fat Transfer to the Breast (up 41%)
  Labiaplasty (up 23%) ← **Sex Related**
  Buttock Lift (up 21%) ←
  Fat Transfer to the Face (up 17%)
  Breast Implant Removal, AKA explantation (up 13%)

- Nonsurgical procedures were up 7% in 2016.

- The nonsurgical procedures that saw the most significant increases in 2016 include:
  Photorejuvenation (up 36%)
  Hyaluronic Acid (up 16%) ⭐ **Image Related**
  Laser Tattoo Removal (up 13%)
  Nonsurgical Skin Tightening (up 12%)
  Botulinum Toxin (up 8%)

  3.5%   7%

- **Injectables overall** (including Belotero, Botox, Dysport, Juvederm Ultra, Juvederm Ultra Plus, Perlane, Poly-L-Lactic acid, Radiesse, Restylane, Voluma, Xeomin, etc.) **saw a 10% increase in 2016.**

- **Fat Transfer to the Breast** (using a patient's own fat) increased in popularity by **41%** in 2016, with more than **25,000** procedures performed.

- **Photorejuvenation** joined the list of Top 5 nonsurgical procedures, with more than **650,000** procedures performed, a 36% increase from 2015.

- **Labiaplasty** continues to be a trend-setting contender in the surgical arena, with **23% more procedures** performed in 2016 than 2015, and with more than **35%** of all plastic surgeons now offering this procedure in their practices.

- **Chemical Peels** became one of the most popular procedures for men this year, joining Botox, Hyaluronic Acid, Hair Removal and Photorejuvenation as their nonsurgical procedures of choice.

- **Buttock Lifts** saw a **21%** increase in 2016, demonstrating that **subtlety continues to be a growing trend.**

Source: American Society for Aesthetic Plastic Surgery

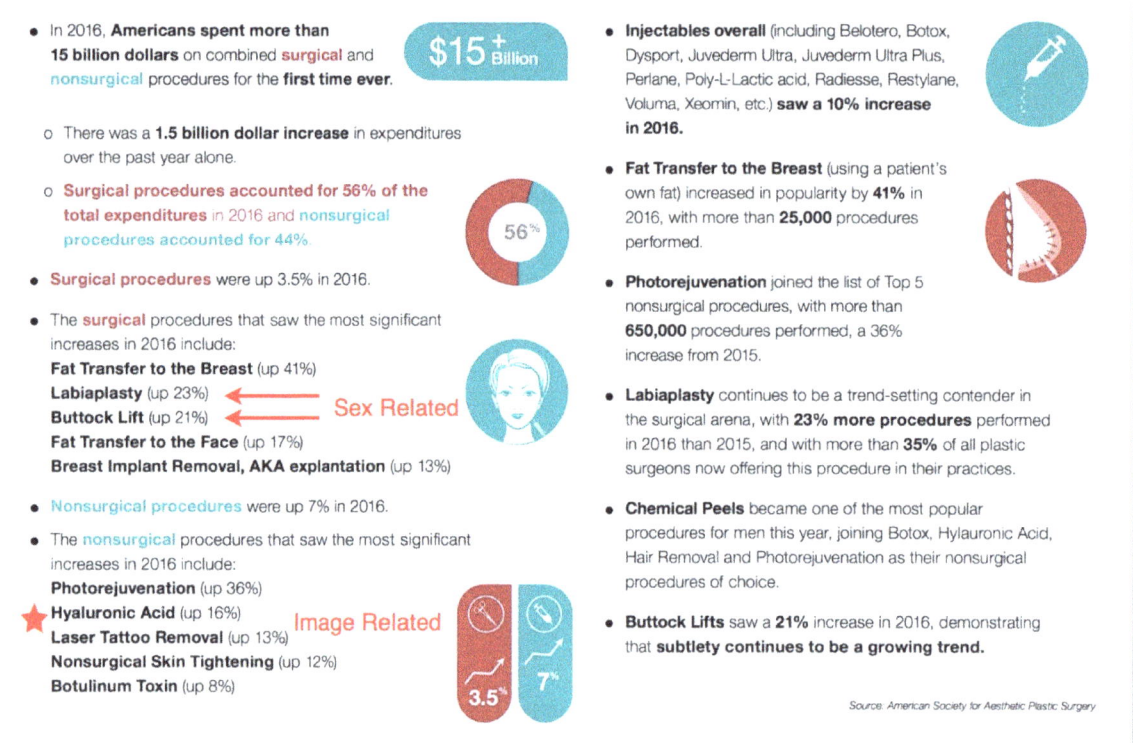

The current author paid special attention to a stock analyst's comments made recently with regard to the stock Snapchat. What does that have to do with a book on the sociobiology of visual image? This is the answer: the stock analyst made a cryptic comment about the image-sharing app when he said this, and I paraphrase: "Who's kidding who, this app is about teenagers and young adults sharing nude photos of themselves."

Look at the data with regard to the dramatic increase in labiaplasty procedures. For those who may not be familiar with this surgical procedure involving female genitalia, female humans have two anatomical features involving what are colloquially termed vaginal "lips." They are the outer "lips" (labia major) and the inner "lips" (labia minor).

Women who seek labiaplasties consider labia minor that protrude to be unattractive. Fuller labia major, as depicted in the before and after photos immediately following, are sometimes judged to be unattractive if they are too full.

The motivation to surgically alter the size and shape of your labia has to be one of the most personal of decisions—or is it?

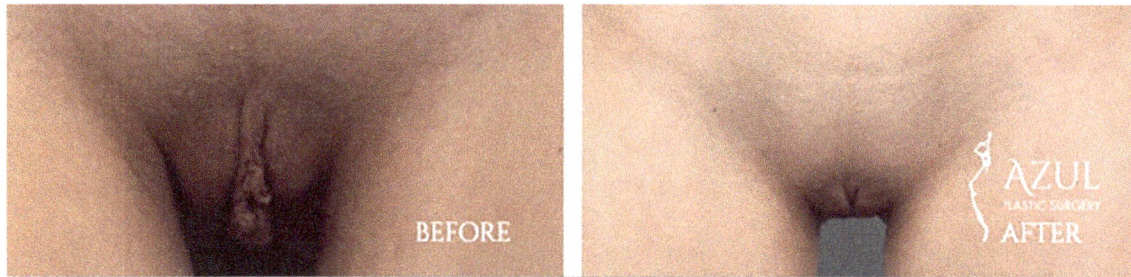

Before and After Results from Labiaplasty Surgery (Photo credit: AZUL Plastic Surgery)

The after photo of this labiaplasty patient has turned back the phenotype clock to a pre-pubescent/pubescent time in the patient's life (note, the patient had a Brazilian wax of her pubic area). Is "turning back the clock" part of the allure of labiaplasty surgery? If it is, what does that say about our culture when grown women want to have genitalia that resemble pubescent or a child's genitalia? Also, what does it say about men who prefer grown women's external genitalia to resemble those of a child or pubescent teen?

The face is always visible. It is logical, therefore, that a person concerned about his or her facial appearance would consider plastic surgery. The labia are not readily visible unless one displays their genitalia to an audience. That audience may be your significant other, consumers of pornography, or social media consumers. Women who are in the business of displaying their genitalia for money are a distinct minority. But women from all walks of life are seeking out labiaplasty surgery. What can account for this increase?

An intriguing catalyst to undergo labiaplasty surgery involves the psychological influence that pornography and social media have had on the general public's body consciousness. Until the advent of social media, girls and women had limited opportunities to compare and contrast their labia with others. The digital revolution changed that because it afforded easy access to pornography, but also, the sharing of intimate photos of oneself is now no more than two or three clicks or finger-swipes away.

Readily available photos of female genitalia provided an opportunity for any one person to make an unfavorable comparison of their own genitalia, compared to the genitalia of those who had photos of their genitalia made and subsequently published.

Is the digital revolution the primary reason that labiaplasties have increased year over year for the last decade? My clinical opinion is yes. Is the worldwide trend toward the sexualization of youth part of the desire to undergo labiaplasty surgery? My clinical opinion is yes. Tampering with time itself is not limited to genitalia; the desire to turn back the clock explains, in large measure, the increase in injectables, laser skin treatments, cosmetics, and the ever-present photo-retouching apps.

Since breast augmentation is the most popular plastic surgery requested by female Americans, let's focus upon that procedure to better understand its connection to sociobiology. It bears repeating: it appears that all of the social engineering efforts designed to amalgamate men and women into one amorphous gender have failed miserably, especially when it comes to the female breast.

Women understand that one of the easiest ways to increase their attractiveness coefficient is to increase their breast size. Researchers like myself draw that conclusion not merely from the plastic surgery data, but from the billions upon billions of dollars that are spent on form-enhancing bras, clothes that accent the breast area, and the amount of breast tissue displayed in public. Yes, breasts continue to function as a primary sexual cue to heterosexual men and lesbians.

What is it about breast size that increases a woman's beauty as measured by her sexual attractiveness? Granted, sexual attractiveness is not technically the same as beauty, but also, the overlap between sexual attractiveness and beauty curves is significant.

Breasts are first and foremost a body structure designed to feed the infant human. A pregnant woman's breasts naturally become larger and firmer than her un-pregnant breasts. Pregnancy and the resultant larger and firmer breasts connote fertility by definition; and fertility and the reproductive act (sex) is the reason there is sexual desire in the first place. I know that this fact is less romantic than mankind's fairytale creations that serve to suppress this unnerving reality, but it is what it is.

Have you ever heard the alliteration: "Big, beautiful breasts?" That alliterative phrase illustrates how big breasts and sexual attraction are, for all intents and purposes, one and the same for many people. Nothing, however, when it comes to the sociobiology of visual image is ever that simple. Beauty is known to connote health and reproductive vigor, which accounts for a large part of mankind's attraction to beauty. Large breasts, absent the beauty of their owner, lose much of their attractiveness factor. Ptotic breasts, though they may be large, can be a source of ridicule because ptotic breasts correlate with infertility, vis-à-vis the "X" variable of age and reproductive vigor.

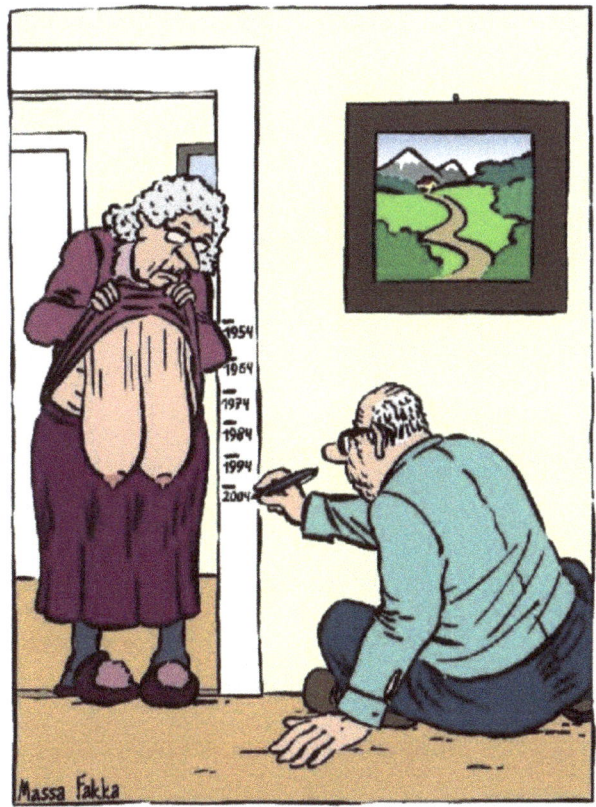

Ptotic Breasts

The shape of the breast can connote youthful vigor and, once the owner of those shapely breasts is impregnated, the size variable comes into play. These dynamics become even more interesting when we examine how beauty interacts with breast size and human behavior.

The vast majority of women tend to misinterpret the attention they receive from men. This is especially true for beautiful women. I repeat, beautiful women make the mistake of concluding that men are looking at *them* when in reality men are looking at *it*, with "it" being their beauty.

Women with naturally large breasts learn, sooner or later, that their breasts garner them attention. Women with average or smaller sized breasts are typically aware of the breast size-attention relationship, and some of them choose, because of their unconscious drive to sexually compete with other women, to increase the size of their breasts. However, what these smaller-breasted women don't realize is that once they obtain larger breasts, and this is especially true for women who choose 400 cc or larger (the equivalent of large C or D cups) breast implants, is this: men instantly become more interested in their breasts than in the person who owns them.

Women who naturally have very large breasts typically resent the fact that men just want them for their breasts, especially if the women in question are also attractive. Women who parlay their large breasts into an image career tend to view their breasts as their "money makers."

Breast augmented women who discover their newfound attraction are often shocked at how quickly they become tired of the fact that men are mostly interested in them for their breasts. Since women tend to not imbue their own breasts with the same sexual symbolism, as do men, women find men's fascination with their breasts to be at best an exploitable male weakness, and at worst a problem so bad that they have their breast implants removed. I have seen this many times. Women with beautiful but large breasts undergo a breast reduction (mastopexy) surgery to squelch their breast fetish suitors.

Of the 290,000 breast implant surgeries performed in 2013, 23,000 women chose to have their implants removed. What percentage of these 23,000 women removed them for the unwanted attention they received is empirically unknown. However, based upon my own research and experience, approximately 33 percent of women develop an ambivalence regarding their decision to have breast augmentation surgery because of how her implants have changed her relationship dynamics with men. A patient's dissatisfaction with her implants, to the point of having them removed, is often times not because the surgeon did a bad job or that she developed problems with the implants themselves; but rather, because the implants increased her beauty-attractiveness coefficient, and these augmented patients came to resent the fact that men were attracted to them *primarily* because of their artificially large breasts.

Whether or not that ambivalence reaches the point where the patient asks for her implants to be removed or replaced with smaller implants depends upon what other attraction characteristics she has. If it is only her breasts, and she has experienced the repeated trauma of only being treated as nothing but a pair of large breasts and not a whole person, the chances of her choosing to have her implants removed is actually quite high.

My clinical experience has informed me that a relatively large number of women undergo breast augmentation surgery to please their boyfriends or husbands. In fact, I have long urged that plastic surgeons be cautious of operating on women whose boyfriends or husbands are the driving force behind the motivation to have breast augmentation surgery. What about when it is a woman in a committed relationship who wants to have much bigger breasts?

I have also bitten my tongue many, many times when I have consulted with women and their husbands/boyfriends, in seemingly committed relationships, who are highly motivated to have breast augmentation surgery when their boyfriend or husband genuinely tell me, "I am happy with her exactly as she is now." I view this situation as a harbinger of bad things to come in their relationship. My clinical experience informs me that when your wife or girlfriend wants bigger breasts when you, the boyfriend or husband, could not care less—watch out!

So, beware of "breast men" if you are a woman. And if you are a "breast man," then be careful of revealing what really turns you on. Of course, this is easier said than done when confronted with a beautiful woman who has undergone breast augmentation surgery.

Ethologist Nickolaas Tinbergen shed some light on the psychodynamics of breast augmentation when he published research on the stickleback fish. Let me remind you skeptical readers that I told you beauty was a multifaceted subject. The female stickleback fish is most attractive to the male of its species when its abdomen swells. Tinbergen wondered what the male stickleback fish would do if presented with a plastic female stickleback designed with an artificially large abdomen.

Tinbergen discovered that when given the choice between the naturally distended abdomen of a fertile and healthy female stickleback fish and an artificial, sterile stickleback that had a cartoon-like large abdomen, the male stickleback fish tends to choose the artificial fish over the natural stickleback, sometimes trying to mate with the dummy fish to the point of hurting itself. Are human men any different? I seriously doubt it. I also wonder what the naturally endowed female stickleback fish would say about the men of her species if she could talk. To be candid, I already know; I've heard it a million times.

MODIFYING YOUR LOOKS IN A BIG WAY

Elective aesthetic plastic surgery has traditionally been viewed as a way to improve an existing set of visual stimuli. I still believe that one of the best compliments plastic surgery patients can receive is when others remark, "You look really good, what have you been doing for yourself?" As surgical technology has improved and as patients have embraced the idea of not just improving, but redesigning their looks, we are beginning to see post-operative patients who are virtually unrecognizable when compared to their preoperative look.

In the past, when plastic surgery resulted in patients who no longer looked like themselves, we have judged such surgeries to be examples of plastic surgery gone awry. Classic examples of surgery gone awry would include the late Michael Jackson's redesigning efforts, along with any number of other famous people who did not necessarily look better after plastic surgery; rather, they looked like a different person altogether.

We have begun to see plastic surgery results that are attractive redesigns, but also make the patient almost unrecognizable from her pre-operative look. When people redesign themselves to the point of being almost unrecognizable, those mercurial qualities that makes us beautiful and, therefore, recognizable and sometimes popular, may become casualties of our redesigning efforts.

A classic example of this redesigning surgical phenomenon is Jennifer Grey. Grey starred in the movie "Dirty Dancing." Grey was easily recognizable for her anterograde nasal bridge and eyelid configuration that gave her a uniquely sexy and mischievous look. Jennifer's facial configuration helped to make her an up-and-coming star with a unique and charming beauty; that is, until she underwent plastic surgery to have her nasal bridge flattened and an upper blepharoplasty that transformed her mischievous and sexy eyes into attractive, but less emotive and identifiable eyes. After her redesigning efforts, Grey's career trajectory flattened.

Jennifer Grey Before and After Surgery (Both Beautiful but Different)

A more recent example of good surgery that resulted in a virtually unrecognizable patient involved the actress Renée Zellwegger. Miss Zellwegger possessed a baby face structure paired with draped eyes that gave her a unique beauty that was highly identifiable; that is, until she redesigned her face with the assistance of her plastic surgeon.

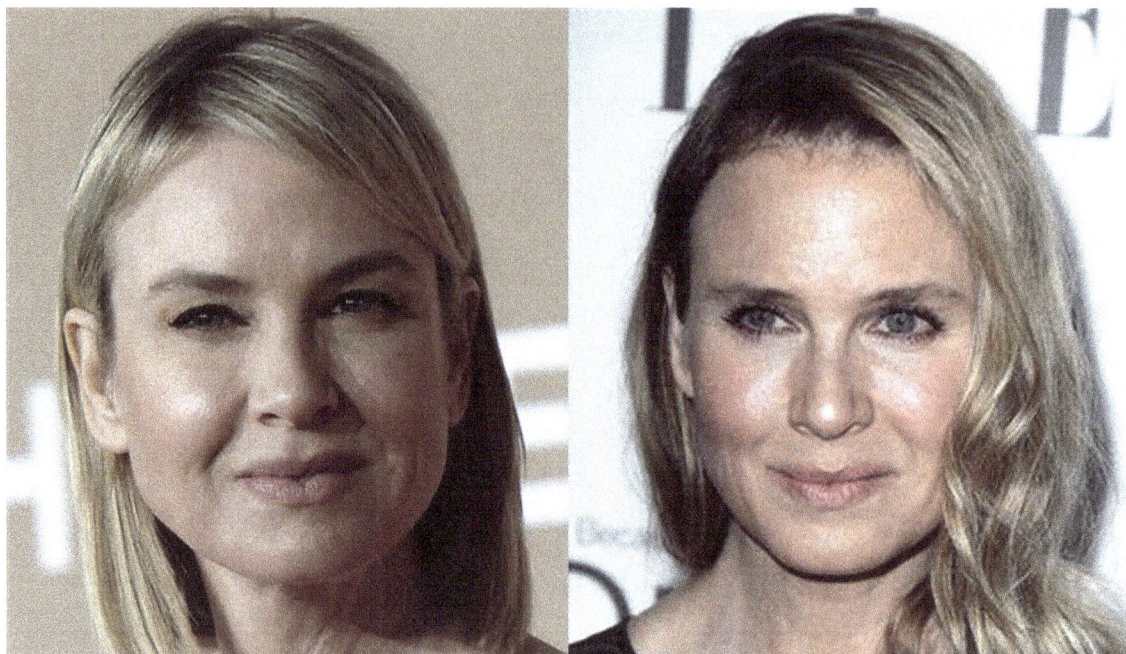

Renee Zellwegger Before   **After Surgery**

Miss Zellwegger's transformation resulted in a person who is attractive but not easily recognizable, when compared to her very popular pre-operative look. Miss Zellwegger had her unique beauty surgically altered and replaced with a different look, not merely a "better" version of the same version.

Beauty is a conglomeration of subtle and oft times mercurial visual cues. Earlier, I identified certain parameters of beauty that, when even slightly modified, often result in a diminution of beauty. But beauty is much more complex than a set of known parameters, even though those parameters provide surgeons valuable frames of reference and guidelines.

As a researcher, I identified what I refer to as "sterile beauty." Sterile beauty serves to illustrate the complexities of beauty. Sterile beauty is beauty that lacks character. It is relatively easy to draw sterile beauty. One of the easiest ways to create sterile beauty is to design a visual image of the face that is mathematically precise in left and right symmetrical features where the proportions between top, middle and bottom thirds are precisely crafted. What you will get if one creates this mathematically precise image is an attractive face, but one that is described as cold or lacking depth. Edgar Allan Poe said: "There is no exquisite beauty... without some strangeness in the proportion." I agree.

What I believe Jennifer Grey's and Renée Zellwegger's plastic surgeons did was to unintentionally modify their patients' ever-so-subtle beauty cues that helped to create their unique brand of beauty, that gave them that special something, or as the French say: "Je nous c'est quoi."

The often nuanced and numerous visual cues that comprise beauty are often so subtle that few can specifically identify them. Adding to their mercurial quality is the fact that such physical attractiveness cues cannot be ascertained, much less understood, unless they are placed within the patient's gestalt facial matrix.

Beauty may be diminished or ruined by either changing the canvas upon which particular facial artifacts or configurations reside, or by ever-so-subtly modifying beauty cues or a configuration of beauty cues from the canvas.

Sometimes, both the canvas and beauty's cues are changed. Ironically, both Jennifer Grey's and Renée Zellwegger's surgeons created a more *technically* perfect facial configuration, but in the process, modified those particular beauty cues and the canvas upon which they resided that gave both women their unique, recognizable and, I might add, popular beauty.

Character-laden or subtle beauty is often unappreciated by its owner. Since attractive women tend to be insecure when it comes to their looks, the tendency is to focus upon those artifacts of beauty that are not only unappreciated, but are also much maligned by their owner. When these people go to their plastic surgeons, the inevitable result is that the surgeon is pre-programmed to identify issues with the canvas and/or beauty cues, sometimes not recognizing that the gestalt of the face is what matters, not the individual tissue artifacts and configurations. This is true even though one or more of those tissue cues may not be perfect.

You'll seldom hear from a surgeon words to the effect, "You may wish to reconsider redesigning your very popular face." I recall speaking those exact words to an extremely popular actress who was considered to be the rival of the late Marilyn Monroe. This particular client did not like a few particular aspects of her face that she recognized as not perfect. And they were not perfect. I pointed out some of Marilyn's "flaws" but pointed out that taken as a whole, each cue working in a nuanced and mysterious way made her fantastically beautiful. She listened to me.

Since many patients these days often want a reengineered look, the plastic surgeon finds that he is sandwiched between a rock and a hard place. If the surgeon is smart enough to treat a character-laden and popular face with great deference, he or she is likely to create a result that makes the owner a younger version of what she already has, while maintaining her canvas and beauty cues.

That genre of result will not please the patient who doesn't like her character-laden, though popular, face. I still to this day have a vivid memory of talking a popular starlet out of a rhinoplasty by making the point that her popularity was something to respect, something she should not tinker with IF she wished to protect her brand.

The common denominator in many patients who fail to appreciate their character-laden beauty is self-loathing. We've already documented that people don't know how they look to others. Moreover, successful people, including actors, tend to downplay the importance of their unique look.

Actors like to think that it is their acting acumen that makes them popular. And while I like to think that is true, casting directors know that acting talent alone is not often sufficient. Casting directors are some of the most talented people when it comes to gauging visual cues and what those cues are likely to arouse and mean to the audience. Look at the following casting director's choices:

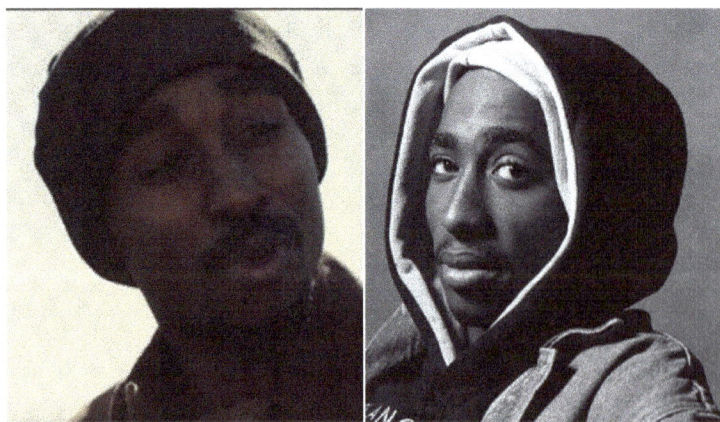

Demetrius Shipp Jr. looks almost identical to Tupac. (Photo Credit: Roadshow)

The Real Eric and Lyle Menendez Brothers with Slain Parents Kitty and Jose

Beverly D'Angelo (Kitty)-Travis Fine (Eric)-Damian Chapa (Lyle)-Edward James Olmos (Jose)
From the 1994 Movie: Menendez: A Killing in Beverly Hills

With rare exceptions, actors and actresses are cast in roles that fit their inherent

visual cues. If you happen to be a popular actor, then you'll do yourself a big favor if you stop before you challenge nature's design efforts with your or your surgeon's remake of your phenotype.

What I have recommended in the past that has worked like a charm is this. I ask the character-laden actor or actress to bring to me a photograph of themselves 15 years younger. I then work out with the surgeon how to achieve that younger look. I encourage the surgeon to treat the faces of their popular patients as if they are walking across a minefield. And, of course, I spend a lot of time educating beautiful people to not pay too much attention to their own self-critical assessments of themselves. When my efforts work, the public never knows about it. Here is but one example of excellent plastic surgery:

Denise Richards

BEAUTY AND SUBSTANCE

Whoever said that beauty is NOT substance, or that those who have beauty lack substance? There is this widespread belief that a person's character, intelligence and related substantive qualities exist in a quasi zero-sum relationship with beauty. Is this true?

First of all, I am the only scientific researcher I know of who has administered state-of-the-art psychometric tests to beautiful women who were participants in regional beauty pageants leading up to the Miss USA Pageant.

Some of my test subjects went on to become state winners, and some went on to participate in the Miss USA and Miss World Pageant Systems. I've administered to these beautiful women IQ tests, personality tests, and psychometrics designed to assess family dynamics. I've also conducted extensive clinical interviews with them. Some of my subjects went on to star in movies and become successful models. Some became doctors, veterinarians and business leaders.

It may come as a surprise to some of my readers that beauty pageant contestants typically have IQs in the bright and higher ranges. In other words, these women are smart. Many, by the way, are excellent students.

Personality wise, they are insecure and surprisingly dependent, both being characteristics of beautiful women in general. This group as a whole is physically coordinated, athletic and healthy. Many have a very good sense of humor, much of it self-deprecating. As a generalization, they are somewhat isolated from their family of origin, no doubt the result of jealousy and envy, secondary to their beauty and the fame that their attractiveness purchased for them in various image related and interpersonal endeavors. Many beautiful women have mothers who are highly competitive with them and/or view their daughters as proof that they, too, are special.

Beauty pageant contestants, at the regional (city) and higher (state) levels, are almost always more attractive than their mothers, sometimes significantly so. Fathers tend to be somewhat distanced from their beautiful daughters, a distance that first manifest when the daughter entered puberty. Sibling rivalry is heightened in this group, especially for same-sex siblings.

Vocationally, beautiful women appear to be both selected for image-related professions and, at the same time, deselected by non-image intensive professions. These facts are true despite the fact that beautiful women certainly have the smarts to enter classic professions like law, medicine, veterinary medicine, business, etc. Many beauty pageant winners have performed community service activities and were, as a group, generous with their time. So where did this notion of beautiful people being dumb and shallow come from?

Average or unattractive people, especially other women, generated this lie. Unattractive men, who were deselected by these women, also joined the intelligence and substance bashing party.

Yes, it would appear, based upon my research, that beautiful women are simply denied their due when it comes to intelligence and substantive worth, mostly by average or unattractive members of their own gender and deselected males. This same dynamic accounts for the promulgation of the stereotype of the dumb blond. It also accounts for the media's tendency to pounce upon the occasional beauty who inadvertently displays her intellectual challenges in public.

Men, as a general rule, are much less competitive with beautiful women when compared to women. Unattractive professional women often cannot control their competitive rage directed at a beautiful woman who attempts to enter her profession, is her co-worker, or lives in her neighborhood. The rage originates in the aggressor's limbic CNS. When the rage reaches the cortical areas of the aggressor's brain, rationalizations and fabricated excuses are used to justify her animus toward her sociobiological (unconscious) rival.

I've written about gorgeous women who became psychiatrists, only to find unattractive or average looking female supervisors who dedicated every waking moment to making the beautiful young doctor's life a living hell. I've seen sitting female judges rake beautiful women lawyers over the coals. Men have their own prejudicial way of interacting with beautiful women, but it typically does not include competition; it is more likely to involve jealousy and issues of control. Is it any wonder that many beautiful women tell me that the only friends they can truly trust are gay men?

The temptation (fame, fortune) to enter an image-related business is often overwhelming for beautiful women. When that temptation is paired with the prejudice that beautiful women receive if they try to break the mold and enter a non-image related business, it is understandable how economic forces alone account for why so many beautiful women end up in image-related activities. The subtle and not-so-subtle pushes and pulls on beautiful women begin early in life.

Beautiful little girls get attention from boys and their male teachers, but are often competed with by their average or unattractive female teachers. I can hear some of you saying, "Adults competing with children?" Yes, beautiful, charming or highly intelligent children often arouse competitiveness in insecure parents, teachers and other adults. The set of visual stimuli we refer to as beauty is hard to hide from would-be competitors. And remember, all of us began life as children, and many adults have vivid memories of their first painful experiences with envy, jealousy and competitiveness.

Studies have documented that most adults can remember the names of their childhood adversaries or tormenters. Similarly, studies show that the vast majority of adults can remember the beautiful little girl or popular boy who got all of the attention in grammar school. Go ahead and take a trip down memory lane; what you come up with may surprise you.

Beauty and substance are paired early on in life, only to be subjected to various pressures to separate as the child matures. One of the first recognizable forced separations occurs when beautiful and intelligent women apply for their first significant job.

If men are doing the hiring, the beautiful woman is more likely to get the job. If a woman is doing the hiring, she is less likely to be hired regardless of her academic achievements or other qualifications. Take note of this profound fact: beautiful women are seldom hired or rejected for substantive reasons, all things being equal.

When it comes to being hired as an on-air presenter, newsreader or on-air meteorologist, beauty can have a distinct advantage. The tendency among male program directors is to hire beautiful AND educated on-air personalities in lieu of average or unattractive-looking AND educated on-air personalities. The vast majority of program directors are men. I have been told numerous times this has to be the rule, because female program directors are "way too competitive with the type of woman we are looking for."

It is as though a beautiful woman's education serves as a rationalization for why beauty was hired in the first place. And this you can be assured of: if the degree is in law, medicine, psychology or business, but the package is beautiful, it will be the beauty that garners viewers, not the academic background. If you are smart enough to have earned a professional degree, you should be smart enough to know that it is your beauty that got you the job as an on-air pundit, but man's defenses are strong and that awareness is not always the case.

The late programming maven Roger Ailes, the man who helped to make FOX News Channel (FNC) the powerhouse that it is, conceived of what came to be known as the "leg chair." As the name implies, this is the chair typically occupied with an attractive, mini-skirt wearing on-air person. Yes, many times these women are smart and educated, but it is called the leg chair and not the IQ chair for a reason.

Leg Chair (Lauren Sivan) Exemplar from FOX Red Eye

Leg Chair (Kimberly Guilfoyle) Exemplar from Fox and Friends

And let me ask you, would you rather watch beauty *and* smart, or simply smart? Your answer will say more about you than anything else. But regardless of how you answered that particular question, the Nielsens are in: people prefer to watch beauty and a little substance over substantial substance without beauty. Given the choice between beauty and substance, in any profession where visual images are its stock in trade, all things being equal, beauty wins.

The assumption readers make when reading a treatment of the beauty substance issue is that there is something wrong when men prefer to watch beauty. The descriptor "wrong" is an example of the phenomenon itself; namely, that beauty does NOT deserve or should not garner the attention that it does. Given the meaning of beauty and how looking at beautiful people and things triggers our limbic systems, nothing could be further from the truth. The fact of the matter is, relatively unattractive men and women are trying to cull or neutralize the competition when they manufacture prejudice against beauty by encouraging guilt over mankind's love of attractive visual stimuli.

BODY PARTS

The visual stimuli we refer to as beauty adheres to Gestalt psychology's fundamental axiom. That axiom states that the whole is greater than the sum of its parts. It is the "package" that warrants the label "beautiful person." Seldom is the label of beauty attached to a person simply because of one body part or area that happens to be beautiful. Though beautiful people adhere to Gestalt's primary axiom, there are differing values ascribed to our body parts depending upon any one individual's tastes and cultural contexts. Nevertheless, certain body parts are highly valenced for most people, regardless of personal tastes or culture, because of archetypal/genotypic blueprints.

We've all heard the descriptors "he's a breast man" or "he's a leg man." When it comes to what women find to be attractive, many women are especially attuned to eyes, jaw lines, or the ratio between a man's shoulders and his hips. For men it is the waist-to-hip ratio. For women, it is the shoulders-to-hip ratio.

To be perfectly clear, both men and women can seldom articulate exactly what catches their attention. This is why researchers like myself have developed clever ways to actually know things even our subjects don't know. Remember, human beings use any number of psychological defense mechanisms to excuse or explain things they do or feel that are either not understood and/or would be disapproved of, IF the true drive or motivation was exposed or revealed. When you use these defense mechanisms, you never fool experts. You may fool others like you, but the one person you almost always fool is yourself.

We have developed ways to identify where people look, how long they look, and the sequence of where their eyes are pointed. This means researchers completely circumvent your opinions and defense mechanisms in order to get at the truth. We can identify patterns of gaze, e.g., what parts of the body are looked at and in what sequence. We can correlate eye gaze with other physiological measurements. For example, we can monitor fluctuations in blood pressure and heart rate when the test subject looks at this or that particular body part. Penile plethysmography and vaginal photoplethysmography are just two scientific methods used to assess sexual responses to visual stimuli.

Some cultures value areas of the body more than other cultures. In Brazil, for example, the derriere is emphasized and has a greater beauty valence than does the derriere in the Netherlands. In fact, the plastic surgery procedure designed to augment, highlight and lift the derriere is commonly named "The Brazilian Butt Lift." In Iran, the eyes and nose have a special significance. This helps to account for the fact that the most popular plastic surgery in Iran is the nose job (rhinoplasty). For example, the eyes you see peeking out of an Iranian burqa may be adorned with exquisite makeup.

Sub-Saharan Africans tend to have an obsessive focus upon derrieres. The same is true for much of Latin America. This obsessive focus upon the derriere appears to manifest quite early, even for black children adopted as infants and raised by Caucasian parents.

White North Americans, including Canadians, appear to have a thing for breasts; whereas African and Arabic peoples, not so much. People throughout Asia, but especially Japan, appear to be in love with relatively large blue eyes. In fact, many younger Japanese girls prefer eyes that look larger, especially when they have their photographs taken. The Japanese call these enhanced photos Purikura Photos.

Classic Purikura Photos

Notwithstanding personal taste and culture, it is the human face that is the most highly visual stimuli-valenced part of the body. No matter where you go on planet earth, a beautiful face can help to compensate for less attractive body parts.

North American women who possess large and shapely breasts often garner the attention of both men and women who have a breast fetish. Women with this particular body part configuration should exercise caution when it comes to men, or women for that matter, who pursue her. Breast fetish suitors are unlikely to declare their fetish. Instead, they will lie and jump through hoops to gain access to the objects of their desire. Women with naturally large and shapely breasts are typically in denial about what she has that attracts men, even more so than women in general who consistently fail to distinguish between the attractiveness of their beauty and the person who possesses the beauty.

Men who possess chiseled jaw lines, have broad shoulders, narrow hips and also have attractive eyes, especially blue eyes, can be irresistible to some women. Women often lose touch with the value of such things as character, intelligence, industry and heterosexual energy when around this genre of man, especially if he has a "cad-like" quality.

Women often convince themselves that a man with this particular body configuration, with less than optimal behavioral characteristics, can be "fixed" or turned into what she wants if she only puts her mind to the project of reengineering him.   This reengineering myth is common among women in love, but really comes into play when women fall in love with a man whose body configuration fits the above description. This configuration is known to trip a woman's primitive biological triggers. Ann and Nancy Wilson of the band "Heart" wrote about this genre of man in their No. 1 hit, "Magic Man."

Cold late night so long ago

When I was not so strong you know

A pretty man came to me

Never seen eyes so blue

I could not run away

It seemed we'd seen each other in a dream

It seemed like he knew me

He looked right through me

"Come on home, girl" he said with a smile

"You don't have to love me yet

Let's get high awhile

But try to understand

Try to understand

Try, try, try to understand

I'm a magic man."

Winter nights we sang in tune

Played inside the months of moon

Never think of never

Let this spell last forever

Summer over passed to fall

Tried to realized it all

Mama says she's a worried

Growing up in a hurry

"Come on home, girl," mama cried on the phone

"Too soon to lose my baby yet, my girl should be at home!"

"But try to understand, try to understand

Try, try, try to understand

He's a magic man, mama

He's a magic man"

Moving on to another body part, girls with long legs have a distinct following. Elvis Presley recorded the very popular song "Long Legged Girl."

All right

I've been thumbin' rides travellin' light

Walked the streets till past midnight

Trampin' roads, trails and lanes

Scaling cliffs fields and plains

Searchin' till the early dawn

For that long legged girl with the short dress on

As noted earlier, the derriere has a huge following. The Brazilian band *Tequileiras do Funk* recorded the very popular song Surra de Bunda, which translates as: Butt Beat. The following are a few select lines from the lyrics of this song:

[O]ur spanking ass !!!

Slams ass , matches the ass !

Knock, knock , knock ! With the ass !!!

Slams ass , matches the ass !

Knock, knock ! With the ass !!!

In clinical psychology, we have identified sexual fetishes that are fixated upon particular body parts. One of the more common fetishes is the foot fetish. It is the foot that holds special meaning to those who have this particular fetish. And though it is unnerving to many people when I ask them, if you had a foot fetish where would you work? The answer is always the same, "Why yes, I'd be a shoe salesman."

I'll go ahead an insert the perfunctory disclaimer at this point that, of course, not ALL shoe salesmen suffer from a foot fetish. Nevertheless, if you do have a foot fetish, working as a shoe salesman in a brick and mortar store would afford you the opportunity to fondle and see huge numbers of naked feet all day long. To these people, it would be like having a job where all you do all day is see and fondle women's breasts.

What follows is a listing of the body part fetishes involved with what we clinical psychologists refer to as Partialism:

Formal name	Common name	Source of arousal
Podophilia	Foot fetish	Foot
Oculophilia	Eye fetish	Eye
Maschalagnia	Armpit fetish	Armpits
Mazophilia	Breast fetish	Breasts
Pygophilia	Butt fetish	Buttocks
Nasophilia	Nose fetish	Nose
Trichophilia	Hair fetish	Hair
Alvinophilia	Belly button fetish	Navel
Alvinolagnia	Belly/Stomach fetish	Belly
Hand fetishism	Hand fetish	Hands
Leg fetishism	Leg fetish	Legs
Lip fetishism	Lip fetish	Lip
Neck fetishism	Neck fetish	Neck
Ear fetishism	Ear fetish	Ears

Recall earlier that I told you I've always conceptualized psychopathology as falling on a continuum, with traits on one end of that continuum and full-blown psychopathology on the other. I told you that we would revisit that issue. Well, now is the time to do that.

You don't have to have a formal diagnosis of Partialism to prefer or be especially attracted to a particular body part or group of body parts. I, for one, am a skin man. I learned that about myself on the first day I attended a face lift surgery during my medical psychology fellowship in plastic surgery. After I scrubbed and entered the surgical theater, I found my beautiful patient comfortably waiting to be placed under anesthesia. About 30 minutes later, as she lay there with her entire face peeled away and draped down below her chin, I asked myself, was my patient still beautiful? "Not without her skin," I answered.

Fast-forward two weeks, and my beautiful patient turned out to be even more beautiful because she now looked younger and refreshed. As I studied her beautiful surgical outcome, I could not get the image of who she is beneath her thin layer of skin out of my mind.

For those of you who are curious, the thickest part of a human's skin in on the soles of our feet and palms. That skin is about 1.5 mm thick. The thinnest skin is on the eyelids. That skin is about .05 mm thick. How does that convert to inches? I say it like this: only 0.001968504 inches of skin stands between beautiful eyes and horror movie eyes. Makes you think, doesn't it? Perhaps you realize, given these facts that you, too, are a skin man/woman.

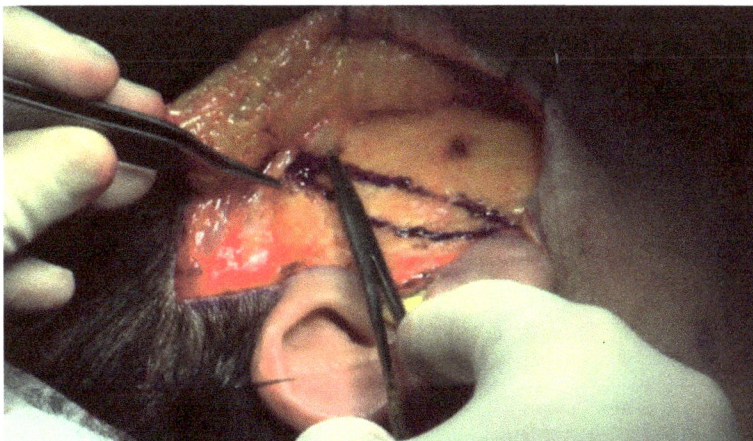

Patient During Facelift Surgery

GUILT

Guilt accompanies beauty as if it were a debilitating tariff on the import of a desirable commodity. If you are beautiful or handsome, you dare not acknowledge it to the public at large. If others make note of your beauty, you'd better be sure to be humble in your response, e.g., "Do you really think I'm pretty? Thank you so much."

Guilt is so consistently and effectively imposed upon beauty, that even when it is obvious that beauty is the sole reason a person has their job, position or popularity, one dare not give voice to that fact. Nowhere is this guilt seen more vividly than in the microcosm of the beauty pageant world. Mind you, it wasn't always this way.

In the period between the 1920s and the early 1960s, women were regaled for their beauty with no apologies needed. Men aspired to marry such women; program directors and ad agencies clamored to retain the services of these women *because* they were beautiful. Girls coming of age in the 1950s wanted to be Miss America, become models or actresses or become spokespersons when they grew up. All of that changed in the 1960s.

The 1960s ushered in a cultural revolution. One big part of the 1960s Cultural Revolution involved the wholesale assault upon beauty. Numerous scientific studies have documented that the most vocal mouthpieces of the 1960s Cultural Revolution were and are, as a whole, comprised of not merely average looking, but frankly unattractive women and men who dedicated themselves to reengineering the West's relatively guilt-free celebration of beauty.

A big part of the 1960s Cultural Revolution included a full frontal assault upon not only beautiful women, but also beauty itself. This meant that the things beautiful women did and do, e.g., participate in beauty pageants, become models, become spokespersons and on-air talent, became targets of ridicule for those interested in promoting societal guilt over beautiful women's inherent sociobiological advantages. I know, that wasn't the social engineers' stated reason for starting this war, but I could not care less about rationalizations when I know the true reasons this war was started and has been waged continuously for over 75 years.

1960s social engineering efforts changed our overt responses to beauty by exponentially increasing the guilt factor associated with it. This had the effect of driving society's love of beauty underground. One of the first manifestations of imposed guilt upon celebrating beauty involved the unspoken requirement that if you are beautiful, you had better engage in acts of contrition for the benefit of those less attractive or less fortunate.

In the early 1980s, a trend began where beauty pageant contestants emphasized their work with disabled children, the deaf, the poor and other welfare-related activities. Contestant's *stated* vocational aspirations transitioned from model, actress or TV journalist to speech pathologist, civil rights lawyer or vague aspirations that included the following taken from actual responses to the question: "If you win Miss _____what would you like to do with your title?" Answers: "I want to do something for poor people." "One day I want to start my own orphanage." "I don't know right now, but I want to help disabled children."

While these aspirations are inherently noble and occasionally did represent a long-held aspiration for these particular contestants, it was clear to me that the genesis of the vast majority of these responses was to serve the function of an act of penitence or contrition designed to wash away the guilt inculcated into these contestants by unattractive women for having committed the sin of being beautiful.

Another trend that occurred in the post-1960s Cultural Revolution was the inclusion of unattractive women as beauty pageant contestants. The spectacle of unattractive girls competing against stunningly beautiful women in a **beauty** pageant created a perfect case study in Group Think distortions. [20]

Everyone knew, including the audience, judges (unless purposefully "placed" to create and promote this social engineering ideology), and pageant directors, that these unattractive contestants served the role of guilt-assuaging tokens. Despite this across-the-board awareness of what was going on, no one said anything due to

---

[20] Janis, Irving L. (1982). *Victims of Groupthink.* Houghton Mifflin.

the Group Think guilt that had been imposed by the PC Police, comprised almost exclusively of unattractive women.

Beauty was deemphasized in almost every way imaginable, though the name of these events remained *beauty* pageants. For example, the swimsuit competition of some regional pageants was replaced with a "sportswear" segment. Despite all of the guilt and social engineering efforts, the best predictor of who would eventually win remained, as it always had been, the person who is most attractive. At least that is the way it *used to be* before ideologically driven social engineers entered the picture.

At the turn of the new millennium, social engineers decided to see how far they could push their anti-beauty agenda. In some cases, patently unattractive women were socially engineered into the winner's circle at the expense of model-perfect women. As social engineers began to openly articulate their condemnation of beauty and the pageants that celebrated it, a sea change transformation took place. Animus against beauty and the cultural celebrations of it morphed into targeted attacks upon identifiable phenotypes. A phenotype is the visible manifestation of a person's DNA, i.e., genotype.

As a preface to this subject matter regarding targeted attacks upon identifiable phenotypes, the reader must come to understand the sociobiological significance and meaning of beauty pageants and related cultural celebrations. Your education, in this regard, will provide to you the supportive foundation to my assertions regarding the nexus between ideologically driven social engineers' genocidal rage and their victims.

Beauty pageants are sociobiological rituals whose provenance links directly to man's reproductive survival as a distinct people (genotypes/culture). Since time immemorial, humans have gathered the most beautiful women (reproductively desirable) in the village to show off their charms to potential suitors (reproductive partners). The archetypal, albeit subliminal, purpose of any beauty pageant is to facilitate the reproductive viability of the tribe (tribes are **always** comprised of a homogeneous genotype and culture). Beauty garners much of its attraction from the fact that genotypic health (reproductive viability and vigor) manifests as phenotypic beauty.   This is true for both men and women.

What follows is a representative sample of various reproductive rituals (modern name "beauty pageants") from across the globe. These include **homogeneous genotypes** of reproductive-age women from each predominant race associated with each geographical area. The following pageant photographs are from pageants that took place within the last three years (2017).

Miss Global Nigeria Winner and Runner-Ups

Miss Arab World Contestants

Miss South Korea

Mr. and Miss World Kenya

Miss World Philippines

Miss World

Please take note of the fact that each tribe's (genotype's) contestants are homogeneous in terms of **predominant genotype**, i.e., race and gender of the geographic area sponsoring the reproductive ritual. In other words, Miss South Korea possesses a Korean genotype, Miss Arab World possesses an Arabic genotype, Miss Nigeria and Miss Kenya both possess a Sub-Saharan African genotype.

> **Since genotype is reflected in phenotypic facial and body configurations, the reader will notice a homogeneous phenotype among the preceding beauty pageant contestants from across the globe.**

The pattern found within tribal rituals, including fertility and/or mating rituals, denoted by a homogeneous genotype among participants found across the globe, has one glaring exception. That exception is a dead giveaway of two things about the sociobiology of visual image. The first is that female beauty is a marker of reproductive vigor in reproductive-age women; and second, competitive rage is both interpersonal AND it is tribal. This means that competitive animus is one-on-

one; that is, directed at a single competitor or target AND it is genotypic, or tribal; that is, directed at an entire genotype/race.

On the interpersonal level, social engineers have been imposing the tariff of guilt on beauty for well over a half-century. But social engineers have now gone so far as to disrupt, with malice aforethought, one particular genotype's (tribe's) reproductive, mating and reproductive fitness rituals. Citing just one example in order to illustrate the principle, ideologically driven social engineers in 2017 went so far as to impose upon pageant directors and judges an average-looking Nigerian woman (Sub-Saharan African genotype) as the winner of a Finnish beauty pageant comprised of model-beautiful Northern European female genotypes. Let's study the data.

According to the CIA World Factbook, over 98 percent of Finland is white. Finnish women possess a phenotype (the visible expression of a person's DNA blueprint) easily recognizable by any expert in the modeling or image business and the more astute or simply honest among the general public.

It is not as though there is no recognizable Scandinavian phenotype. In fact, Scandinavian female genotypes are overrepresented in the modeling business. Peruse the following photos of Scandinavian models. As you look at these photos, you'll recognize the classic "Scandinavian look" that is the phenotype produced by the Northern European genotype:

Top Ten Finalists in the Miss Helsinki, Finland 2017 Beauty Pageant

Winner of the Miss Helsinki, Finland 2017 Beauty Pageant, Nigerian Sephora Ikalaba

Keeping in mind the sociobiological provenance of beauty pageants, what ideologically driven social engineers executed in Helsinki in 2017 was a bold assault upon the Scandinavian genotype. This bold assault was designed to destroy the cohesiveness of the targeted tribe's reproductive ritual by imposing an average-looking disparate tribe's genotype into the winner's circle. It just so happens that empirically, in all cases studied across the globe, only Caucasian tribal rituals have been subjected to tampering and interference. Ideologically driven social engineers' logic underlying this assault makes perfect sociobiological genocidal sense: If you can destroy/disrupt a competing tribe's reproductive rituals, you can carry out a Fabian-style genocide of that tribe's unique genotypic fingerprint.

Fabian-style acts of sociobiological aggression are always shrouded within egalitarian shrouds, e.g., diversity, multiculturalism or some other socially engineered false "ideal" that was fabricated to serve as protective cover for the underlying genocidal rage that defines this genre of tribal aggression against targeted genotypes.

If diversity were the true motivation behind these social engineering efforts, then why not insist upon diversity for ALL tribal cultural rituals, which are homogeneous in terms of genotype, and not limit your assault upon Caucasian people and their rituals?

After all, *when placed within the world's populations, white people are a distinct minority (1 in 6 people).*

In reality, not only do ideologically driven social engineers single out Caucasian reproductive rituals of beauty for disruption/eradication, but also,

All other (non white) tribal rituals that openly and proudly reject diversity, i.e., the participation of other racial/genotypic groups, are actually promoted, encouraged AND PROTECTED by ideologically driven social engineers so that they remain homogeneous in terms of genotype, race and culture.

For example, in May of 2017 a biracial (African and Caucasian) woman won the Miss Black University of Texas Beauty Pageant. Immediately upon winning, Rachael Malonson was viciously attacked by ideologically driven social engineers, their minions, and black people that she was, and I quote: "Not Black Enough." Imagine if the winner of the 2017 Miss Helsinki pageant had been criticized as "not being white enough."

> "[B]ut several people on social media expressed disappointment in the fraternity for selecting a winner who they say does not represent the black community because she does not look black. Some Twitter trolls even outright asked what her race was and questioned her blackness, even after being told she was biracial.

> Malonson actually anticipated skeptical responses to her participation. "I wasn't sure if I would even place in the pageant because I wasn't sure they would think I was 'black enough'," she told USA TODAY College. She decided to enter anyway. "I chose to do the pageant to gain a deeper inner confidence before I graduate, while breaking

stereotypes that black people or mixed-race people have to look a certain way," she said." [21]

DJ~
@DJ2779

Follow

She's black?

12:50 PM – 1 May 2017

4

Soufflé Steve @nyleswashington   2 May
Replying to @dantreselove and 5 others
No what?

D'Antrese
@dantreselove

Follow

Let me ask a better question ... is she black ?

8:42 PM – 2 May 2017

11

chocolate drop 🍫 ✨
@_ColeWorldShwty

Follow

she's clearly the lightest, damn near white looking "black woman" & she won over the others. quit playing dumb guys. it looks sketchy

10:05 AM – 3 May 2017

39   216

Jameeda Rucker
@_JRPR_

Follow

they should pulled out the brown paper bag test 😂 🤷🏽‍♀️

9:11 AM – 3 May 2017 · Cincinnati, OH

3   7

---

[21] USA Today College. Reported by: Briana Stone. May 3, 2017.

Ideological social engineers recognize that beauty, and the cultural pageants that celebrate beauty, represent an Achilles' heel that can be subverted in order to undermine cultural and genotypic homogeneity, as well as reproductive viability, in a targeted group of people.

It is an empirically demonstrable fact that ONLY Caucasian tribal rituals are under subversive attack by ideologically driven social engineers. The fact that visual image (beauty) resides at the heart of these social engineering attacks, which target white culture and white genotypes and their accompanying phenotypic visual stimuli, reinforces visual image's deep significance in almost all matters of the human experience, including genocidal rage and tribal aggression.

I will give the reader an iron-clad guarantee that Miss Nigeria, Miss Kenya or Miss Djibouti will never be won by a blond and blue-eyed woman, or that the winner of Miss Beijing will never be won by a redhead from Ireland. I will also guarantee to the reader that ideologically driven social engineers will never impose upon the Miss Djibouti pageant a winner who happens to be an average-looking Caucasian girl. Why the reverse is promoted as "good" by ideologically driven social engineers, and accepted by so-called "social justice warriors," when it comes ONLY to Caucasian genotypes and culture, should tell you everything you need to know about what is truly going on from a sociobiological perspective.[22]

Grand scale social engineering is almost always paired with uniquely personal competitiveness and hatred of beauty. For example, social engineers made it less troublesome for people to shroud their appreciation and love of beauty by providing to them a protective labyrinth of rationalizations, compensatory justifications, and various acts of contrition and penitence.

The more that unattractive women browbeat men to give up their love of beauty (for public consumption) and, instead, embrace unattractive women like them, the more the underlying competitive dynamic became obvious. Simply put:

---

[22] Readers indoctrinated by those same ideologically driven social engineers have been conditioned to **reflexively** reject and be **repulsed** by ANYONE who would dare to expose what truly motivates ideologically driven social engineers' targeted assaults. I ask the reader who is in search of FACTS and not ideologically defined litmus tests to pause to consider the truths presented herein.

Unattractive women and racialist social engineers have ganged up on and found a way to use guilt as a cudgel to beat up the beautiful and more attractive competition, going so far as to eradicate their reproductive rituals.

Ideologically driven social engineers' assaults upon beauty have become so loud, it is deafening. Ironically, the more unattractive women promoted their social engineering virtues at the expense of mankind's inherent love of beauty, the less attractive they and their cause célèbre became.

Ever since social engineers began their assault upon beauty, absolutely nothing fundamental has changed on the interpersonal level of social interaction/attraction. Beautiful women remained the most popular girls at the party, the most likely to be the television presenter, the girlfriend or wife of the rich and successful man, the object of desire by successful lesbians in show business, or the ideal object of study for plastic and cosmetic surgeons, artists, photographers and selfie-takers.

When it comes to men, progressive women still flock to see male matinee idols that are attractive. To even suggest that beauty, either for men or women, can be socially engineered away is a fool's errand born of abject ignorance of the true sociobiological nature of beauty and its cultural celebrations.

While unattractive women were burning their bras and wearing granny panties, Victoria's Secret designers were busy creating ever more frilly, sexy and beautiful bras, panties, slips and nightwear for beautiful women, or those who wanted to be beautiful.

The more men were told that they were shallow and stupid for lusting after beautiful women, the better they became at ignoring the shrill and unattractive voices in their midst. *As if men needed any more help, feminists became—and this was no easy trick to pull off—even more unattractive*.

To be sure, many attractive women were successfully co-opted, if not persuaded that they were "bad," by unattractive women. Attractive women were inculcated with guilt and a pernicious form of mind control in order to encourage them to join the unpopular ranks of unattractive women who dedicated their lives to becoming unmarried spinsters in service to the revolution they had been programmed to promote.

On the other hand, grand scale, e.g., tribal ideologically driven engineering efforts have been more successful. The reason they have been successful is because the targets of their aggression have bought, hook, line and sinker, their protective covers of multiculturalism and diversity. Also, ideologically driven social engineers' targeted genotype and culture have been made to hate who and what they "think" they are.

If targets (victims) come to realize that multiculturalism and diversity are reserved ONLY for their genotype (tribe), but for no other tribal entity (genotype), one would hope that the truth of what is motivating ideologically driven social engineers' crimes against your tribe of humanity would become obvious. I'm not holding my breath. The reader should never forget this fact:

> ***Sociobiological competition lay at the heart of ideologically driven social engineers' assaults upon beauty and the cultural events that celebrate it. Distinct tribal entities financing, designing and carrying out these assaults upon YOUR tribe are vehemently protective of their OWN genotype and use laws, culture and social norms to enforce THEIR genotypic homogeneity.***

Beautiful women who naturally possess a reproductive advantage became the target of unattractive, unhappy and dissatisfied women, along with progressive men, who executed an eradication campaign upon beauty and reproductively fit people and cultures. Where did this animus originate? It originated from the deepest recesses of deselected men and women's primordial instincts.

When you boil it all down, ideologically driven social engineers and their feminist co-conspirators intended to eradicate the pretty girls who represented their competition in the life and death struggle that is sociobiology, i.e., reproductive viability. They even went so far as to disrupt/corrupt their targets' DNA while protecting their own unique DNA. You've heard it said many times, but it is worth repeating: You can't fool Mother Nature, but it appears you can hurt and sometimes kill Mother Nature's creations.

SPACE MEN AND BEAUTY

One of the more fascinating aspects of beauty is that the more you learn, the more you begin to realize that beauty is anything but skin deep. When researchers looked into the nature of beauty, they began by cataloguing those physical properties that comprised the variables in the complex equation that defines beauty within any given time and culture. I've discussed some of these variables in this book, including shifts in taste when it comes to beauty over the decades and across cultures. But there is a more fascinating aspect to the general subject of changing beauty preferences over time.

I have noted that if we expand our changing tastes in visual image from across decades and centuries, to comparisons across the millennia, what we find is a shift in preference for a body type and features that mirrors our species' known evolutionary trajectory.

We've all seen the sequence of pictures of man where in his earliest form he is covered in hair, stooped over and very ape-like, progressing left to right in a sequence of more upright, less hairy, paler-skinned images that evolved into modern man.

Modern man looks at images of his predecessors and judges those images to be unattractive. You would be hard pressed to find someone who looks at our primordial ancestors and remarks, "My, what a handsome fellow." Exaggerated supra-orbital ridges (those bony structure above the eye sockets upon which eyebrows grow) are experienced by modern man as primitive or unattractive.

When you see a modern man with a strong supra-orbital ridge, deep eye sockets, pronounced occipital ridges, short and slanted forehead and extensive body hair, the vast majority of modern men and women tend to judge that person to be unattractive-looking, even if the people making the observations have no knowledge of anthropology or evolution. The more a modern person shares the features of our ancient ancestors, the less attractive we judge that person to be. Scientific judgments about beauty, if you haven't already learned, are made irrespective of wardrobe or hairstyles when it comes to evolutionary tastes.

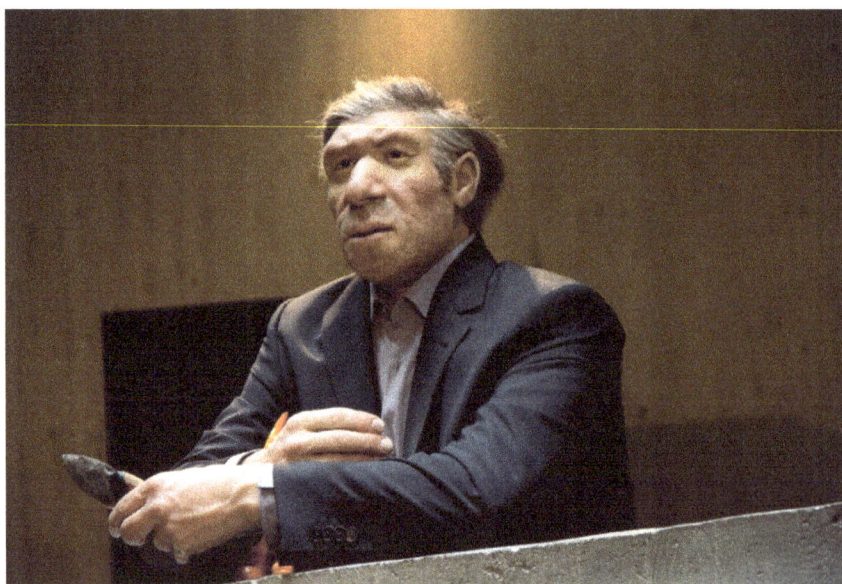

A Primitive Looking Man Wearing Modern Formal Clothes

INTERNATIONAL MALE

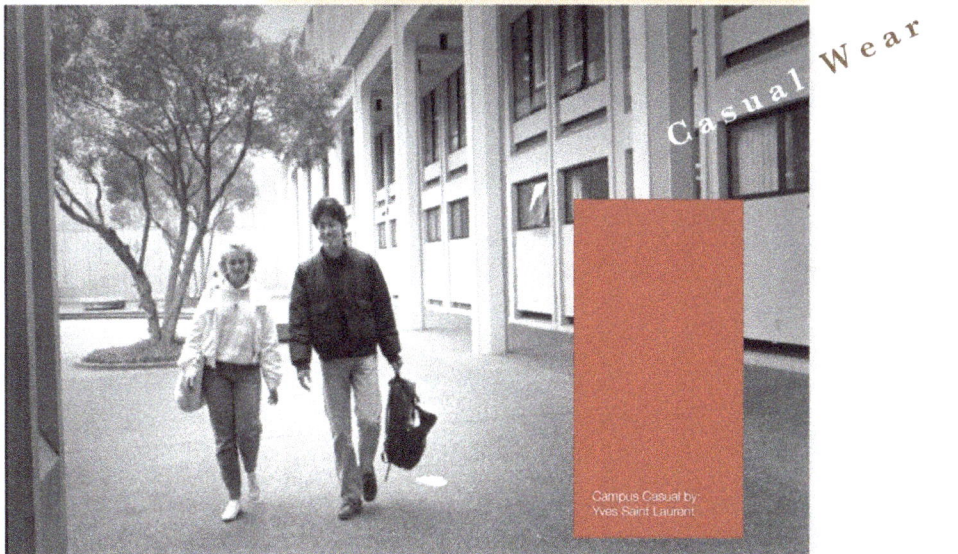

Formal Wear

JOHNNY CARSON
suits and their models
Late night fits.

Casual Wear

Campus Casual by
Yves Saint Laurent.

Modern Man Wearing Formal and Informal Clothes

Using our primitive ancestors as a reference point appears to confirm our species' preference for modern-day body and facial configurations, but what about our future preferences?

The future of beauty preference can be predicted based upon known trend lines. In other words, the human body is changing in known ways, as are our preferences; and if one were to extrapolate moving forward along that trend line, we can approximate how we will look in the future.

The next point I want to make is fascinating, because it opens the door to the notion that we human beings have a species-specific archetypal body type ideal rooted in our genes. To further explore these ideas, let's investigate the human body type trend line over the next thousand years.

Human bodies are becoming taller, thinner and less robust skeletally. Our limbs are becoming more elongated and wispy. Our heads are becoming larger, especially our cranium's frontal lobes. Our eyes are becoming larger and more pronounced as the bony orbits within which our eyes are situated are becoming shallower. Our ears are becoming smaller, and our necks are becoming longer and thinner. We are losing our body hair and our skin's pigment is becoming less intense and lighter.

Permit me to illustrate species-specific somatotype trend lines by using the Barbie Doll as a reference visual image for this phenomenon in trend lines. What follows is a comparison of the current nominal body type of the female human with the body type dimensions of Barbie:

21ˢᵗ Century Representative Female Homo Sapiens

Average woman's height is 5'4″
Their weight is approx. 140 lbs.
They wear a size 14 dress
Their bust is between 36″ and 37″ (B cup)
Their waist is between 30″ and 34″
Their hips average between 40″ and 42″
Their shoe size is estimated to be between 8.5 and 9.5

BARBIE

Barbie's height would be 7'2″

Her weight would be 101 lbs.
She would wear a size 4 dress
Her bust would be 39″ (FF cup)
Her waist would be 19″ (same as her head)
Her hips would be 33″
Her shoe size would be a 5

The popularity of the Barbie Doll has been framed by progressive observers as a social construct created by advertisers, white hegemony and misogynistic men. The criticisms of Barbie's body dimensions are motivated, in part, by how average women are allegedly made to feel when they compare themselves to Barbie. Racialists' criticisms of Barbie are because of her Caucasian phenotypic facial and body characteristics.

Because many other racial phenotypes do not look anything like Barbie's facial and body characteristics, Barbie has been the object of intense criticism by ideologically driven social engineers. Rather than create dolls with non-Caucasian facial characteristics, which would be the most logical and parsimonious solution to this alleged manifestation of social injustice, racialists insist upon eradicating Barbie's futuristic body type and facial characteristics.

Consider for a moment that if Neanderthal men had created a doll that had the facial and body characteristics of modern women, the proportional differences between Neanderthal women and modern-day women would approximate the differences between modern-day women and Barbie.

Why is the Barbie Doll popular?" Is she popular because she was made popular by advertisers, white hegemonists and misogynistic men? Is she popular because of what ideologically driven social engineers call "white privilege?" If Barbie's popularity was manufactured out of thin air, as ideologically driven social engineers and progressive social engineers boldly assert, it would also have to be true that toy makers could socially engineer a Neanderthal-looking doll that would be just as popular as Barbie, if she were promoted as the ideal to the same degree and intensity as Barbie was promoted. Would such a heavily promoted Neanderthal doll be as popular as Barbie? The answer is no, of course not.

A Neanderthal doll would not be as popular, no matter how much advertising and social engineering effort was put into creating her popularity. This is because we humans appear (regardless of our tribe's phenotypic characteristics) to have a wired-in preference for body types that are further along (more advanced) on the evolutionary trend line. This includes the finer facial features and less intense melanin characteristics of some existent human genotypes.

Barbie may be modern man's vision of where we are going, not where we are or have been in our evolutionary trek. Barbie's popularity may derive in large measure from the fact that modern man finds an evolutionarily advanced look to be attractive, including smaller and more refined facial characteristics, which just happen to coincide with more refined facial characteristics found in some specific genotypes, e.g., the Danakil Tribe of Africa. [23]

Among all other Sub-Saharan Africans, Danakil women are disproportionately represented in the modeling business and are more popular (considered to be more beautiful) among **BOTH** indigenous Sub-Saharan Africans and Sub-Saharan African Americans than are average-looking Sub-Saharan African women.

---

[23] The Danakil Depression is the northern part of the Afar Triangle or Afar Depression in Ethiopia, a geological depression that has resulted from the presence of three tectonic plates in the Horn of Africa.

Danakil Woman (Note her finer and more refined facial characteristics and less intense melanin concentrations)

An Average Looking Sub-Saharan African Woman with Child

Modern man's behavior, as reflected in his self-directed body engineering efforts, suggests that man has demonstrated a preference for a more evolutionarily advanced body type. Yes, there are exceptions, but the rule remains. Many of man's self-body image-engineering efforts mimic an evolutionarily advanced look. What about so-called modern-day primitive peoples who are isolated from the media entertainment complex? What do they prefer when they engineer their bodies and face?

The women of the Kayan Lahwi tribe in Burma lengthen their necks. They tend to lighten their skin, because lighter shades are considered to be more beautiful when compared to the extremes of melanin concentrations, i.e., darker skin.

The exception to this is tanning as is popular among people with lighter skin types. However, tanning has much less to do with a preference for a darker skin color than it does with the skin homogenization effect that tanning has on lighter-skinned people. Tanning makes translucent skin, with its visible veins and mottled look, a more uniform and reflective canvas, i.e., more beautiful. This is the same visual effect that can be achieved with makeup.

And speaking of skin color, I want to share with my readers some of my previous work on this subject. The following is an excerpt from my textbook *Awakening Beauty: An illustrated look at mankind's love and hatred of beauty.*

"Skin Color and Stereotype

In part one of a study conducted by researchers at Tufts University, college-age subjects from the greater Boston area were shown pictures of light- and dark-skinned black men and women. Subjects were then asked to match up a series of neutral statements with the photographs. In part two of the study, researchers asked subjects to list the traits they believed were commonly associated with light and dark-skinned black people. One of the more clever aspects of this study was a design that isolated skin tone from other physical cues, such as hair texture, lip fullness or nose width.

What the researchers found was that both black and white people ascribed differing descriptive adjectives to African-Americans according to the lightness or darkness of their skin. In addition to being judged as more attractive, lighter-skinned black people were judged to be more intelligent and more likely to possess wealth. Conversely, darker-skinned black people were judged to have less money, be tougher or more likely to be a criminal.

Dr. Keith Maddox, the lead author of the study who happens to be of African-American descent, told a reporter for the Boston Globe, shortly after his work was published, that the subject area of his research has been thought of as divisive and explosive and something we shouldn't ever talk about in public. Dr. Maddox went on to say: "If you don't try to understand it, then things will never get better."[24]

---

[24] The current author could not agree more with this particular statement by Dr. Maddox.

By way of background, many of the students who served as the test subjects of this study were born in the 1980s. Many come from middle-class and privileged environments.

The college age young men and women who made up the test subjects of this research, if they are anything like the average white and black American, love Michael Jordan, Barry Bonds, Eddie Murphy, Chris Rock and are fans of numerous black rap artists. Furthermore, much to the chagrin of purists in the hip-hop culture, many of the white young men in the test subject pool, if they are anything like white young men across America, so admire hip-hop culture that they have adopted many of the mannerisms, speech, dress and attitudes associated with hip-hop culture. Add to these sociological data the fact that every person on Earth seems to ascribe positive adjectives to that which is beautiful.

This phenomenon is so widespread that Dion, Berscheid and Walster entitled their landmark research in the early 1970s: "What is Beautiful is Good." People from all cultural backgrounds tend to ascribe more positive traits to good looks. In 1941, a researcher by the name of Monihan studied the perception of criminals by experienced workers in the criminal justice system. She concluded that even battle-hardened social workers find it hard to believe that beautiful women can be guilty of a crime, noting that people tend to think of criminals in terms of abnormality in appearance.

If lighter-skinned people are perceived as more attractive and less threatening, as the Tufts study data suggest, the question becomes, then, why are darker-skinned people less attractive, and is there a beauty connection?

A more sociobiological-beauty related interpretation of Dr. Maddox's research findings involves the visuo-spatial image processing associated with light-absorbing images versus light reflecting images. Darker-skinned people lose some of their inherent three-dimensional beauty cues because of the loss of contrast-effect, which is part and parcel to darker skin. And with regard to the "menacing" adjectives ascribed to darker-skinned people, e.g., "aggressive" and "tough," many of the facial cues, which inform others of kinder intent and motive, are lost to the observer who looks at a darker-skinned person. In the modeling business, "tanned" models are often described as "hard."

People with mid-range skin tones are easier to see when compared to very dark skinned people. Mid-range skin tones reflect more light, and the shading within their particular range of "lightness" affords a more vibrant contrast effect, when compared to a darker-skinned person.

If contrast effect and beauty cues account for some of the negative adjectives ascribed to darker-skinned people, wouldn't it follow that very light-skinned people would also suffer a similar prejudice? As a matter of fact, very light-skinned people are stereotyped with negative adjectives, not unlike their darker-skinned counterparts. And like their darker-skinned brothers and sisters, very light-skinned people, as for example albino populations, also lose much of their facial cue information related to beauty and those facial designs which inform others of a kinder intent, because of the loss of contrast which comes from a very light skin tone. Dr. Vail Reese, a San Francisco dermatologist, told "Wired News" about his research on the subject of albinism. Dr. Reese has chronicled the negative portrayal of albino men and women in movies. Matrix: Reloaded, aka Matrix 2: Reloaded, included villains known as "The Twins." Other more recent movies that have portrayed albinos as evil characters include Star Wars: Episode II, Attack of the Clones; Die Another Day; Blade 2; and Star Trek: Nemesis. The movie The Time Machine included an entire race of albino creatures called "Morlocks." According to Dr. Reese, "audiences recognize it very quickly. They've learned that if you see a character with albinism in the movie, it's going to be an evil character. It's become part of the film language, for better or worse." (SEE PHOTO CREDITS OF ALBINISM IN THE MOVIES)

Assessment of overall health, an important social cue to others related to reproductive fitness and personal survival, is another factor related to man's attention to skin tone. A culture's reaction to skin color may have a sociobiological basis. On the Western Coast of Africa, Darwin wrote in his work Descent of Man that the natives admire a darker skin more than one of a lighter tint. Their horror of whiteness may be attributed as partly to their thinking it a sign of ill health.

This assessment of health by equating it with various skin tones is absolutely accurate, both in Africa and the Western World. "Paleness" may mean low blood sugar, hypotension, disease or the blanching that frequently occurs just before a person vomits. "Green behind the gills" is a colloquialism that connotes poor health and unattractiveness as judged from skin tone and color.

"Uber" from: *The Time Machine* (Photo Credit: Andrew Cooper, Warner Brothers)

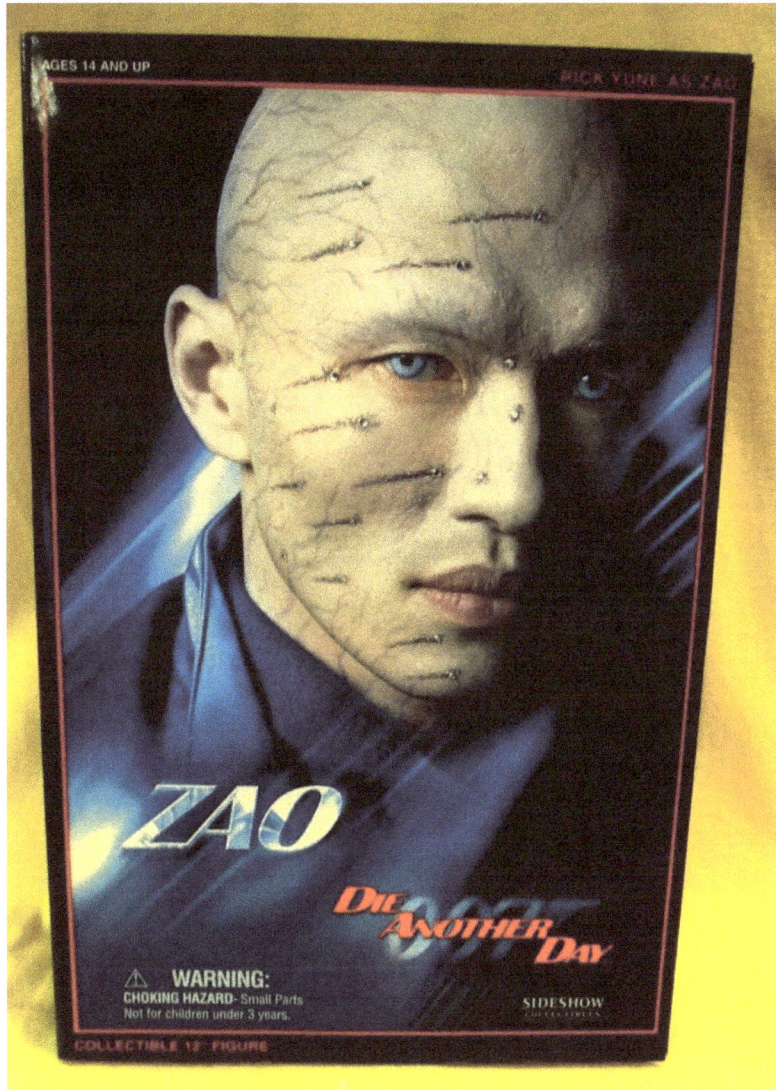

"Zao" from Die Another Day (Photo Credit: MGM Studios)

"The Twins" from Matrix 2: Reloaded (Photo Credit: Warner Brothers)

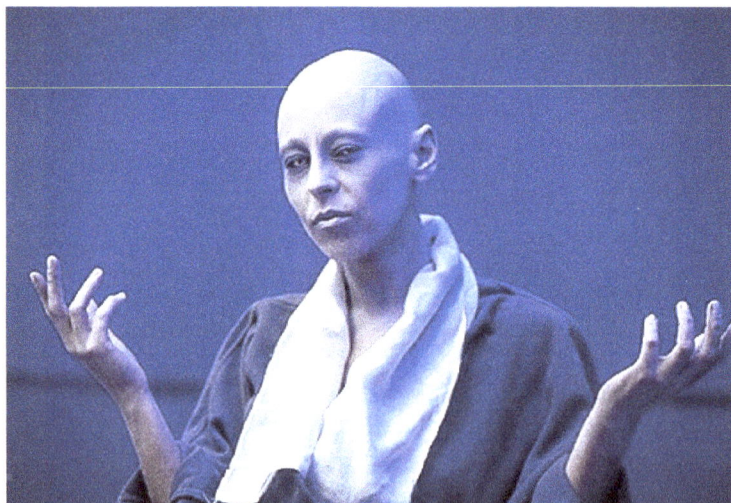

"Sly Moore" from Star Wars: Episode II, Attack of the Clones (Photo Credit: Lucas Films)

Preferences for a particular skin tone and color may have little or nothing to do with deep-seated sociological processes, e.g., racial prejudice. For example, fair-skinned white people who "tan" are never assumed to be motivated by some unconscious sociological motive related to race, e.g., white people who tan are so attracted to black culture that they want to share their skin tone. We do not ascribe a psychodynamic motive of identification to white people who tan, which would explain their desire to tan as an unconscious desire to identify with oppressed and negatively stereotyped darker-skinned people. No. white people who tan simply want to look good. Researchers understand that by tanning a moderate amount, fair-skinned people increase the homogeneity of skin tone and even out blue and ashen shades, which are unattractive to the eye. Similarly, black people who even out their skin tone or increase their luminosity and homogeneity by lightening their skin should not be understood as engaged in anything other than the quest for a marker of beauty that may have little or no relationship to cultural norms. It should be noted that, for example, when Eldoquin-Forte, a prescription skin lightener, is applied to both lighter and darker skin, its primary effect is to even out skin tone by "lifting" the dark spots that are the inevitable consequence of aging, exposure to sunlight and irregular pigmentation patterns in the human melanin complement. Thus, so-called skin-lightening creams are more aptly thought of as skin-homogenization creams. Good skin care, particularly skin care that increases homogeneity and clarity in skin tone, is associated with a dramatic increase in beauty. In fact, skin care, with particular reference to creating a healthy skin tone and color, offers the greatest "bang for the buck" when it comes to increasing attractiveness. When black people apply moisturizing lotion in order to avoid the ashen hues which occur when some darker skin becomes dry, they are not unconsciously or consciously trying to avoid becoming more like some of their Anglo

brethren, i.e., lighter, they simply want to look good. [25]

Perhaps now the reader has a better understanding of the complexities of skin color and how ideologically driven social engineers, fueled by their ideology and woeful ignorance of sociobiological issues, set about to re-engineer the natural world according to their unfounded and, to be gracious, confidently stupid and genocidal motivated social engineering efforts.

As human beings seek their evolutionary body type path, we remove our hair with shavers, depilatories, waxes and lasers. Tightly knit curly hair is straightened. We go on diets, exercise and choose the foods we eat in order to become thinner. We universally use makeup, surgery and ocular implants in order to increase the size of our eyes. All races lighten their skin in one way or another. Do these self-engineered modal changes remind you of anything?

Why aren't these body image engineering efforts reversed? That is, instead of spending billions on weight loss regimens, where are the weight gainer businesses and boutique foods on our grocery store shelves that promise weight gain (excluding body builders, which is a totally separate issue), more facial hair on women, stouter and thicker legs, more pronounced supra-orbital ridges, smaller eyes and thicker body hair? Why don't women brag about their size twelve shoe instead of tending to lie about having a smaller shoe size? Why don't men and women promote body hair growth so that the advertising slogan of "hairier is better" is as common as hair removal ads? The answer is NOT because of Caucasian hegemony or some other manifestation of cultural oppression.

It is NOT a coincidence that the evolution of our species' body type is moving in a direction that is reflected in our self-directed body engineering efforts and fabricated ideals as reflected in our dolls, our video heroines and Photoshopped images. Astronomer Carl Sagan was famous for remarking that we human beings are made of "star stuff." Perhaps we are, and somewhere in our collective unconscious, we know it.

---

[25] Napoleon, Anthony (2003). Awakening Beauty: An illustrated look at mankind's love and hatred of beauty. VBW Publishing, College Station, Texas.

MACHO MEN AND FEMME FATALES

In the animal kingdom, it is relatively easy to ascertain gender by studying the gestalt visual image of the animal in question. As a generalization, males tend to be more colorful and flamboyant than are females. Ethologists tell us that males tend to have dominant combs, horns, colors, sounds, and other visual and auditory cues that are used to both attract the female of their species and communicate dominance to other males. To cite just one example, female birds appear to be attracted to the most colorful and flamboyant male. This is true because of the attraction coefficient of genetic viability. Readers have most likely heard these concepts before, but what do they really mean?

Predators tend to target potential prey because, in part, they are easier to see. If a colorful male is able to be colorful AND survive, then that male's genes are probably viable. Females tend to be less colorful and flamboyant, because drabness has a survival advantage. Again, thanks to the behavior of predators and the immutable law of natural selection. What about us humans?

What men find to be attractive, according to the logic of sociobiology, is a healthy, fertile and sexually receptive female. Those qualities are disproportionately found in compassionate and visually stimulating females. Underlying this logic is the powerful unconscious drive on the part of men and women to pass on their genes. Females, according to sociobiologists, are motivated to be impregnated by the most viable male. This logic is pristine and persuasive, until one factors into the equation that mankind has developed ways to circumvent the connection between sexuality and reproduction. To say it another way, **people have learned to "game the system."**

Viability is a term that is susceptible to confirmation bias. Insensitive men almost always believe they are "viable." The "love 'em and leave 'em" crowd does the same thing. Of course, confirmation bias reasoning doesn't make your biases any more valid. Impregnating a female is the easy part. Sticking around to protect, support, nurture and help your family constitutes the operational definition of a genetically viable male.

Women have often learned this wisdom the hard way because many men, especially those who have been acculturated by a culturally narcissistic zeitgeist, are simply failures at anything BUT the simple act of reproduction, and this is because when it comes to impregnating the female of the species, the bar is set so low. Numerous studies have confirmed that women impregnated and NOT protected and supported by the father of her child is THE recipe for early death, future trouble and a life of poverty for BOTH the female and her offspring.

Humans have socially engineered away the pristine logic of thousands of years of time-tested species promulgation that insured the selection of the fittest. Not only has man learned to disconnect sex from reproduction, thus changing our own species' survival prospects, man has learned to feign sexual vigor and receptivity by manipulating his presentation cues.

Man has removed two lynch pins from his species' survival paradigm. He has disconnected the sex drive from reproduction, and he has learned to feign genetic viability. And how did mankind do this? He learned to short-circuit the link between sex within a committed relationship and making babies, and he learned to fabricate the way he presents himself to a mating partner. In other words, **deselected males learned to lie by manipulating visual cues and feigning character**.

I would be remiss to not address the fact that books on sociobiology, especially those written in a two-dimensional manner for popular consumption, when mixed with deselected males' confirmation bias, are used to justify male predatory, selfish and failure-laden behaviors. I addressed some of these, e.g., "love 'em and leave 'em" and exploiting women's defects when it comes to the ability to discern substance from posing.

As you saw in the opening chapter, young men have learned to rent an expensive car for a day in order to fool girls who are engaged in similarly manipulative game played by the males. Think of the gold digger genre of female as merely the other side ("tails") of the same coin, where on the "heads" side of the coin we find the duplicitous male who is out to engage in a sex act with an unsuspecting victim.

This book should stand as a testament against everything these poser males and gold digger females do and stand for. Welfare roles, aid to dependent children, and government food subsidies are ALL the end result of irresponsible and failed men and women who lack character, who are a drain on culture and society as a whole.

If there is any practical lesson I want to emphasize in this chapter, it is this: This book should help the victims of gold diggers AND the duplicitous failed male to be able to spot these racketeers before they can cause harm and create havoc in your life. So what are some of the visual cues used to trick unsuspecting potential mating partners?

Men use the visual cue of facial hair to communicate "maleness" to women. Men typically have little or no awareness of why they grow facial hair; they just think they grow a moustache, goatee or beard because it makes them look better. The reader should always keep in mind that social conformity is a primary motivator to adopt any visual cue IF that visual cue is popularized vis-à-vis the media using seemingly attractive role models.

The kind of moustache a man prefers changes his look, and that look tells you a lot about the man. The bigger and bushier the moustache, the more there is an underlying need on the part of the man to compensate for his lack of perceived maleness.

The particular style of facial hair can reveal a man's underlying gender identity. In my research I have found that gay men, for example, typically prefer a more pronounced moustache when compared to heterosexual men. Facial hair can also inform us about socio-cultural identification and personality type.

Men who prefer facial hair that must be coiffed daily tend to be image conscious and highly judgmental. Men who wear their facial hair in idiosyncratic patterns, e.g., "pork chop" sideburns or handlebar and waxed moustaches, tend to view the world in an equally idiosyncratic fashion.

Full beards have traditionally been associated with religious men and/or men who tend to have something to hide (not that those two categories are mutually exclusive). Full beards may be intended to hide a personality defect or may simply be intended to cover a self-judged unattractive face. An unkempt full beard may be a sign of an antisocial personality, mental illness, or may serve as a rejection of conformity and the rules and conventions of society.

Men who dress in Western wear who are not from a location where such dress is the norm, tend to identify with a rugged lifestyle that is the exact opposite of who they really are. Cowboy hats, pointy boots, large belt buckles, handlebar moustaches, etc. are all compensatory alterations to a man's visual image.

Metrosexual men embrace their feminine side and are known to alter their looks in ways that tend to make them look pretty, but not masculine in the traditional sense. Metrosexual men tend to be into fashion and are rapacious consumers of skin and hair care products, as well as image-conscious accessories. It should come as no surprise that metrosexual men tend to have been raised by single mothers who raised them as imperfect females.

If I use the term "macho man" most of you will immediately picture in your mind's eye a stereotypical image. Macho men often look and behave like peacocks. They tend to strut about, display their feathers and defend their "territory." Keep in mind that peacocks engage in much of their stereotypical strutting behavior for the benefit of other male peacocks. It is the same for human peacocks.

Macho men tend to wear attire that reflects their *imagined* self-image. They may have a moustache, wear shoes with a prominent heel, unbutton their shirts to expose their chest, tend to wear gold chains and pinky rings, and often wear a belt with a prominent buckle.

Leather jackets are a mainstay in the macho man's wardrobe. If I had a nickel for every man of short stature with a high-pitched voice who wears a black leather jacket, I'd be…well, I'd have a hell of a lot of nickels. Since macho male behavior is primarily for the benefit of other men, this begs the question: do women like macho men? The answer to that question is certain segments of the female population appear to be attracted to macho men. Who are these women?

The female analog to the macho man is the femme fatale, and it is this genre of woman who engages the macho man. Femme fatales tend to dress to the "nines" and are coiffed to perfection, using makeup and hairstyles that are eye-catching. Short skirts, tight pants, high-heeled shoes, bright red lipstick and perfume that can be detected a mile away collectively represent the calling card of the femme fatale. Many of these women are "teasers" in that they appear to thoroughly enjoy attracting, then frustrating the advances of her macho man analog in the pseudo mating ritual that characterizes her playground or killing field.

Femme fatales are masculine in their primary psychological identity, with most of these women having identified with their fathers during their developmental years. Femme fatales' masculine psychological identity is hidden because it is most often wrapped in a caricature of femininity. It is that caricature of femininity that is so very attractive to the macho man who is, like the femme fatale, a caricature of maleness. Underneath the macho man façade resides a little boy who was addicted to approval from his mother. Macho men are often victims of an unresolved Oedipal Complex. Yes, many macho men were momma's boys who were the apple of their mother's eye while having never earned the respect and acceptance from their father, which often resulted in a love/hate relationship with their father.

Macho men are not attracted to the femme fatale's polar opposite, the pretty feminine girl. This is because this genre of demure woman does not enhance his maleness, as he is implored to demonstrate to his male friends. The pretty feminine girl may be somewhat attracted to the macho man, but should she get to know him, she soon learns that she was attracted to a mirage.

Macho men want women who serve the same function as the new golf club they show off to their friends at the golf course. If the macho man's wife or girlfriend enhances his ego when he displays her to his envious male friends, underscore the term envious, then she has served her primary purpose in his life.

Imagine the psychodynamics underlying the matchup of the macho man and the femme fatale. Friday and Saturday nights all over the world are dress-up nights where femme fatales and macho men display their wares in a faux mating ritual that is doomed to fail, unlike its more successful versions in the animal kingdom.

The primary reason the human version of this mating ritual fails is rooted in how easily we humans can be tricked by faux visual cues that speak to our limbic system, but not to the higher cognitive centers of our brains.

Forensic psychology has long understood that it is very common for malintent to dress itself in non-threatening attire. The wolf in sheep's clothing has worked for thousands of years because we human beings judge others by their stereotypical visual cues and the archetypal characters they play in their make-believe movie.

Pusillanimous men can and often do dress like big bad cowboys, and cold, angry women often dress like warm and feminine beauties. Think about that the next time a pair of high-heeled shoes and big red lips, or a big belt buckle accompanied by a moustache, comes strutting into the bar.

FACIAL DOMINANCE

One of the more interesting and profound findings in the visual image research literature is related to how certain facial features predict success and rank in the military. Mazur and Miller (1996) made a career studying facial dominance. Read what Keating, Mazur and Segall wrote about facial dominance:

> *"Some stimulus faces were consistently rated dominant, independent of their expression, while other faces were consistently rated submissive. Thus, facial dominance was judged reliably across cultures. Some people always looked dominant, others always submissive."* (Keating, Mazur and Segall, et al. 1981) [26]

As far as the definition of facial dominance, the following is a very brief overview of what that term means:

> *"What do dominant faces look like? Everyone knows because anyone can sort portraits on this basis, but facial dominance seems to be a gestalt concept, difficult to describe in simple terms. Faces identified as dominant are more likely to be handsome -- with striking exceptions, to be muscular, to have prominent as opposed to weak chins, and to have heavy brow ridges with deep set eyes. Submissive faces are often round or narrow, with ears "sticking out," while dominant faces are oval or rectangular with close-set ears (Mazur, et al. 1984). The descriptor "babyfaced" (Zebrowitz, et al. 1993) presumably refers to a submissive appearance."*

All of these preliminary data are important, because it turns out that facial dominance is an excellent predictor of which young military cadets will rise to the highest ranks over time. In other words:

---

[26] Keating, Caroline F.; Mazur, Allan; Segall, Marshall H.; Cysneiros, Paulo G.; Kilbride, Janet E.; Leahy, Peter; Divale, William T.; Komin, Suntaree; Thurman, Blake; Wirsing, Rolf. (1981), *Culture and the perception of social dominance from facial expression.* Journal of Personality and Social Psychology, Vol 40(4), Apr 1981, 615-626.

"If you knew nothing else except the young cadet's facial dominance scores, you could predict who would be the higher ranked military officers far off in the future."

Here is what the authors wrote about their findings:

"[H]owever, to our surprise, facial dominance -- measured from cadet portraits taken 20+ years earlier -- significantly predicted promotion to the highest ranks -- to the various levels of general officer. Having a dominant face was an advantage in reaching the top." (Mazur and Mueller 1996)

Actor Sir Michael Caine has thought a lot about visual image, including facial dominance. I make mention of Michael's opinions here, because the research on visual image supports his most insightful observations.

"The first thing is, if you are a man with a very small head, or a woman with a very big one... you will never see a romantic couple on the screen, as a success, where the man's head is smaller than the woman's. The woman's head must always be smaller, but you don't notice it. If you saw it, you wouldn't even notice it, you'd just know that you didn't like this couple. If you can see your nostrils from straight on, you'll never do it. If you can see your gums as you're speaking, above your teeth, you'll never do it. If, in the normal relaxed posture of your face, you can see the whites above your eyes, you will never become a leading lady or man, because it's disconcerting."

Larger faces and heads are more common among movie stars and our most successful presidents and military leaders. As with all matters related to visual image, the question arises whether or not there is a substantive basis for man's tendency to choose leaders who possess dominant faces and larger heads?

The answer is yes, there is a substantive reason and it is related, at least in part, to testosterone. Testosterone is an androgen that is found in exponentially larger concentrations in men (300 ng/dl) when compared to women (30 ng/dl). Testosterone is the chemical that is behind the transition of pre-pubescent boys and their higher voices and softer faces into baritone-speaking and hairy-faced young men.

Testosterone changes the human face during our developmental years in ways that are known to anyone who can distinguish men from women. It appears that stronger jaws, larger faces, more dominant brow lines and occipital ridges are visible cues of an underlying tendency toward dominance and viability.

When Mueller and Mazur, et al. documented the link between dominance, facial cues and military rank, what they were really documenting was the connection between testosterone and other male genotypic characteristics that are predicate behaviors and traits that promote success in the military.

Mueller and Mazur's facial dominance studies apply to ANY person vying for a position of leadership or success in any competitive endeavor. The current author offers for a point of reference a man I personally know and have worked with. General Mike Neil's visual image is exactly what Mueller and Mazur wrote about when they discussed facial dominance. After his illustrious military career, General Neil went on to lead efforts to help those less fortunate.

General Mike Neil

General Mike Neil

I want to throw cold water on an almost innumerable number of pop-psychology books (almost always written by lay people) who stupidly, but confidently, assert that they can teach small-faced, small-jawed *and* ambitious men to become a truly dominant male by changing their attitudes or mindsets. Ignorance of the biology of true dominance, not poser or mindset dominance, has spawned a cottage industry for males who crave to be something they are not. If these books teach anything, it is how to lie in order to fool those who lack discernment.

Dominance is not learning to be overbearing, physically threatening, or adopting the behavior of a totem animal. Military leaders who are truly dominant understand that authority is given, not taken from his troops. True authority is never the end result of employing gimmicks and pop-psychology to the management of people. Dominance means being an anchor of strength during tough times or when the chips are down, NOT the common man definition, which is more similar to a WWE wrestling star who struts around the ring in a faux display of dominance.

Dominance is patience, honesty, compassion the protection of those weaker and those in need. Only someone who is genuinely dominant can present a visual image that may appear to be submissive and vulnerable, but is anything but that.

Beta males who crave to be alpha males never, ever comprehend the distinctions I have made here.

Alpha or Dominant Males in Action

Alpha or Dominant Male in Action

Alpha or Dominant Male in Action

Research by David Perret and Anthony Little, conducted at St. Andrews University in the U.K., reminds us that you can see the evidence of testosterone when you compare the faces of young boys and women. Those faces are similar, as are their voices, until testosterone kicks in. According to the researchers, masculine faces are sexy because they are dominant faces.

Research done at The University of Canterbury in New Zealand by Satoshi Kanazawa postulated that testosterone is the driving force behind both intellectual achievement and criminal activity. Kanazawa studied the biographies of 280 scientists. He compared age-related patterns of achievement in scientists, jazz musicians, painters and authors, and found that they matched the age-related patterns of criminal activity in convicted felons, all of which is related to testosterone and facial dominance.

Mazur, Miller, Seagall, Zebrowitz, Perret, Little, Kanazawa, et al. have proven that:

All the attitude and mindset manipulation in the world cannot compensate for the biological reality that defines sociobiological truths that actually define dominance. Never forget that sociobiological truth resides under all that attitudinal and mindset bullshit that appears to be so popular these days.

Politicians in the modern era are more likely to be successful if they are telegenic and have a beautiful speaking voice. The evolution of the modern era mirrors that of media technology. A quantum leap in the history of the modern era occurred in 1960 when the Kennedy/Nixon Presidential debates were televised live.

What makes a candidate telegenic? Facial dominance paired with nonverbal behavior that conveys genuine empathy, warmth and confidence. "I feel your pain" are merely words, absent President Clinton's nonverbal behavior that feigned those words may be true. "Mr. Gorbachev, tear down this wall," would be a hollow command were it not for President Reagan's resolve as communicated by his voice and facial structure and expressions of dominance.

When it comes to women, beauty sets up a dynamic that tends to exaggerate any perceived disparity between image and substance. In other words, if you are pretty, you'd better be smart. If you are a woman who is model beautiful, your candidacy puts you at risk to not be taken seriously. If you are patently unattractive, the probability of being rejected by the voting public is exponentially higher unless your district is disproportionately comprised of people similar to you. Since most politicians are average to slightly unattractive looking, the contrast effect between beauty and patently unattractive candidates rarely occurs as it did in 1960.

Washington D.C. has been described as the place unattractive people go in lieu of Hollywood to be popular and be "stars."

2008 Presidential Vice President candidate Sarah Palin is a classic example of the political substance/beauty dynamic. Though not model beautiful, Palin was more attractive than the majority of her female political peers. Women in the media, who were themselves image conscious and considered telegenic, and who also opposed Palin's politics, mercilessly pounced on her gaps in knowledge and perceived intellectual deficits, both of which were exaggerated by her detractors because of the way she looked.

Besides static features, I am particularly interested in nonverbal communication. Legendary leaders of the modern era have possessed the entire package, so to speak. Not only are their faces dominant and their voices authoritative, but also, their nonverbal behavior communicates confidence and warmth, a relatively rare combination. I make note of the modern era because earlier generations were not instantly connected to the real-time images and sounds of would-be leaders. For example, it took several days before the news of Abraham Lincoln's assassination reached the far Western states.

And speaking of Abraham Lincoln, scholar Harold Holzer has studied Lincoln's nonverbal characteristics. Lincoln was killed 12 years before Edison invented the phonograph, so no one alive has ever heard Lincoln's voice. Nevertheless, much is known about Lincoln's nonverbal behavior. Smithsonian magazine did a review of Holzer's work:

"When Holzer was researching his 2004 book Lincoln at Cooper Union, he noticed an interesting consistency in the accounts of those who attended Lincoln's speaking tour in February and March 1860. "They all seem to say, for the first ten minutes I couldn't believe the way he looked, the way he sounded, his accent. But after ten minutes, the flash of his eyes, the ease of his presentation overcame all doubts, and I was enraptured," says Holzer. "I am paraphrasing, but there is ten minutes of saying, what the heck is that, and then all of a sudden it's the ideas that supersede whatever flaws there are." Lincoln's voice needed a little time to warm up, and Holzer refers to this ten-minute mark as the "magical moment when the voice fell into gear." He recalls a critic saying something to this effect about Katharine Hepburn's similarly startling voice: "When she begins to talk, you wonder why anyone would talk like that. But by the time the second act begins, you wonder why everyone doesn't talk like that." Says Holzer: "It's that combination of

gesture, mannerism and unusual timbre of voice that really original

people have. It takes a little bit to get used to."[27]

Two points are worth noting. The first point is that Lincoln was unattractive, and the second point is that his voice was not the mellifluous baritone we have come to appreciate in our most popular modern era's political leaders. So how did Lincoln make it to the top of the political game, given his presentation deficits? It was his uniqueness, as conveyed by his nonverbal communication, along with his brilliant mind, that made up for his voice and looks.

There was also one other factor. Know what that factor is? There was no media entertainment complex back in Lincoln's day. Whether Lincoln could make it in today's media-intensive environment is debatable, given that one single gaff recorded by the media can ruin a presidential candidate's chances overnight, to wit:

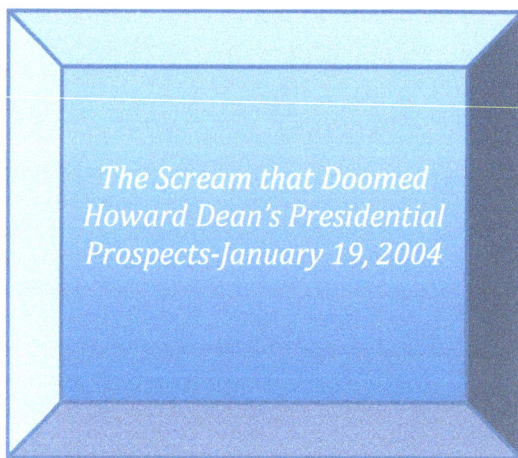

The Scream that Doomed Howard Dean's Presidential Prospects-January 19, 2004

https://youtu.be/l6i-gYRAwM0

All of this brings us back to the beauty/visual media conundrum. The beauty and visual media conundrum is best understood by the issue I raised earlier regarding Abraham Lincoln's chances of becoming president, were he to have lived

---

[27] Smithsonian Magazine (2011), *Ask an Expert: What Did Abraham Lincoln's Voice Sound Like?* By: Megan Gambino.

in the modern era. In 1960, the people who listened to the Kennedy/Nixon presidential debate on the radio concluded that Nixon had won the debate. On the other hand, those who viewed AND heard the presidential debate on television concluded that Kennedy had won the debate.

Seldom do we have such a clear-cut and real life study that compares the impact of substantive content vs. static and kinetic visual image. The modern era emphasizes visual image over substance by a process of conditioning that begins during infancy. Digital natives spend approximately 8-10 hours of each day looking at various screens, the most popular of which is their smart phone's screen. Attention spans of commercial television watchers are curiously correlated to the time between commercial breaks and are known to be limited to minutes, not hours.

Not only are digital natives screen addicted, but also, the images consumed on those screens are comprised of novel, repulsive, sexual, selfies, or a combination of all of the above. Not only are leaders judged by their visual image, but also, it appears that leaders may be little more than their visual image linked to palatable sound bites that are careful to not tax an attention span, and frustration tolerance that is remarkably deficient when compared to Lincoln's generation.

ACTING

Acting is a descriptive term that refers to the management of how a person presents themselves to others. Actors manage their presentation based upon how they and their director want that presentation to be interpreted by an audience. Presentation is transmitted across three channels: visual, auditory and content.

The visual element is comprised of two channels. One channel is the static component. As the name implies, the static channel is comprised of visual stimuli that do not move, e.g., the shape of your face or body type. The other component of the visual channel refers to the changing configurations of our faces and bodies, e.g., our posture and our facial expressions. These are comprised of kinetic visual stimuli, aka, non-verbal behavior.

The auditory component of presentation is comprised of all the musical qualities of sound, e.g., pitch, rhythm, and phrasing of the voice. Content refers to what ideas and concepts we convey to others by and through the words we use.

If you become very good at manipulating your presentation, you may become a professional actor; that is, you may be a person who is paid to look and sound like a particular character or persona. Screenwriters provide the content for actors, who are left to focus upon manipulating their looks and sounds according to how their director wants them to look and sound. Impressionists are masters at capturing the auditory and visual dynamics of famous people. For example, check out the brilliant impressionist Frank Caliendo as he performs his impressions of former Presidents George W. Bush and William J. Clinton.

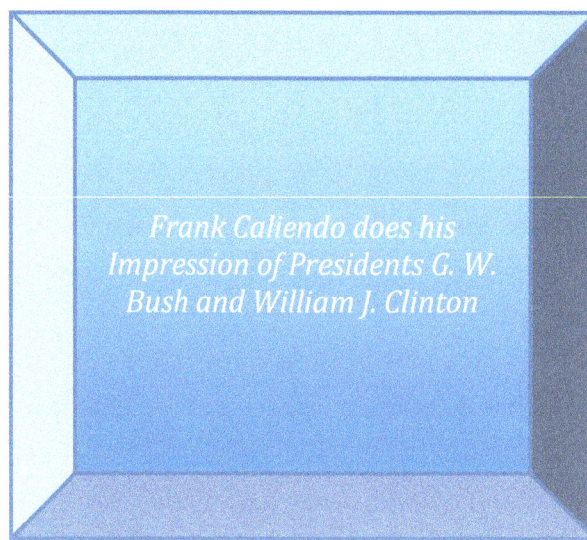

Frank Caliendo does his Impression of Presidents G. W. Bush and William J. Clinton

https://youtu.be/XK1ekhovFeU

When Shakespeare wrote "all the world is a stage," he was editorializing about how all of us passively or actively manipulate our presentation in order to influence others. Passively managed presentations can be referred to as "by chance" presentations. Passive presentations are those visual, auditory and content variables you inherited, usually through osmosis and genetic inheritance, from your immediate environment and your parents. Active management presentations are the visual, auditory and content we choose to manipulate and present to others.

We are acting—that is, managing our presentation—when we engage in behaviors that make us appear more attractive than we would otherwise be. We are also being disingenuous in some sense. And since we humans instinctively prefer genuineness instead of disingenuousness, a moral conflict arises when we manage our presentation.

For example, those people who critically editorialize about the use of makeup; that is, the widespread practice of manipulating our visual presentation using makeup, often use the word "fakeup." People who pretend to be warm and helpful are sometimes pejoratively referred to as "phonies." People who wear high-heeled shoes, color their hair, whiten their teeth, enlarge their breasts, wax their bodies, have plastic surgery, etc., are managing their visual presentation. They are, according to critics, being disingenuous. Is that true? Are these people being disingenuous?

In some sense, people who actively manage their presentation care more about what other people think than those who do little or nothing to manage their presentation. In that respect:

Criticisms directed at those who actively manage their presentation are misplaced. Shouldn't one's criticisms be directed at those who don't give a damn about what you think of them?

Some professional actors possess presentation dynamics that are naturally attractive, regardless of the roles they play. This attraction dynamic may be rooted in static and/or kinetic beauty, charisma, facial dominance, quirkiness, or most likely a combination of all of these variables.

The late actor John Wayne, for example, was always "John Wayne" regardless of the role he played. Wayne's facial dominance, and the charisma that flowed from it, helped to make him successful. The same could be said for Ronald Reagan. No matter what role he "played," including his most famous role, President of the United States, he was always the presentation package we knew as Ronald Reagan. Reagan was charismatic, warm, and affable. He also possessed a mellifluous voice. Similarly, President Kennedy possessed a statically powerful face with an accompanying nonverbal that conveyed a sense of humor and wit. He was also smart, well educated and brave, all of which created a stunning charisma.

The interaction between static beauty and nonverbal behavior can be synergistic, as in the case of Reagan and JKF, or the relationship between static image and nonverbal behavior may be contradictory, or they may nullify one another, as in the case of Hillary Rodham Clinton. The relationship between static and kinetic beauty can and often do have fascinating implications.

For example, 2012 Presidential candidate Mitt Romney was more attractive than his opponent President Barack Obama on the dimension of static beauty. But when it came to nonverbal behavior, Romney came across as cold and distant when compared to the warmth and fluidity conveyed by Barack Obama. Romney's nonverbal characteristics, despite his static beauty, resulted in the moniker "Mitt-Bot."

A famous book on acting is entitled: *Acting is Believing.* [28] The title reflects the author's thesis that the only true way to look and sound the part is to actually become or believe you are the part and/or the person you are playing. In acting circles, this has come to be known as *The Method Approach* to acting.

The idea behind method acting is that the actor who purposefully manipulates the 46 muscles in his face, his body kinetics, and his voice to convey a character, is seldom as successful as the actor who comes to delude himself that he IS a particular character in this or that REAL situation.

Fake smiles are symptomatic of bad acting, but also, fake smiles may be indicative of a person who is so genuine that they cannot fake a smile effectively. In this sense, a fake smile may be the behavior of a genuine person who despises having to put on a happy face to please others. Again, I told you at the outset that visual stimuli were more complex than you ever imagined.

When an actor successfully uses The Method Approach and actually becomes the character he has been cast to play, sometimes he or she cannot undo the

---

[28] McGaw, Charles, Stilson, Kenneth L., Clark, Larry D. (2011), *Acting is Believing.* Cengage Learning.

character transformation. Among actors, this inability to undo the blending of self with the character is termed "submersion."

It is said that when Faye Dunaway played Joan Crawford in the movie "Mommy Dearest," she submerged with her character and actually "became" Joan Crawford for a number of months after the completion of the filming.

The late, legendary actor Sir Lawrence Olivier perfected his acting craft as few others had done. Olivier used both The Method and The Technique Approaches to acting. The Technique Approach is pure acting, in that the actor pretends to look and sound like his character. So what do method and technique actors think of one another?

One of the more interesting stories (some say apocryphal) told in Hollywood involves an interaction between Sir Lawrence Olivier and Dustin Hoffman on the set of the movie *Marathon Man.* One scene required Hoffman to look as though his character had been awake for three days straight. True to his method training, Hoffman actually kept himself awake for 3 days to genuinely become the sleep-deprived character. Upon learning of this, Olivier purportedly said to Hoffman, "Try acting, my dear boy, it is so much easier."

One of the more intriguing aspects of those individuals who have presentation talent, but have no comprehension of the words they speak, is the fact that when people realize talented actors often have no idea about the profound meaning of the words they act out, that realization comes as a shock. Actors without a script who voice opinions on this or that subject often run the risk of bursting the bubble of intelligence, strength, cleverness, facility and sexiness **projected** upon them by the audience. What follows is a classic example of a talented fire and brimstone child preacher (actor) exposed as having no comprehension of the words he so persuasively spoke:

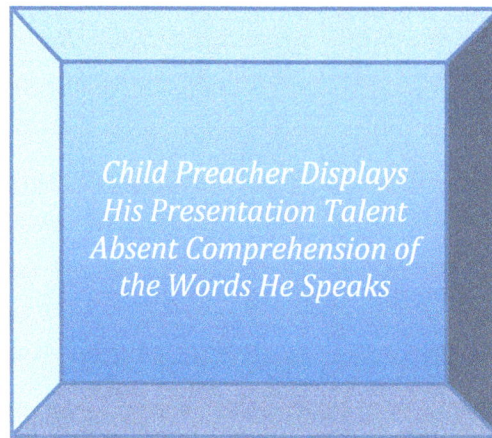

Child Preacher Displays His Presentation Talent Absent Comprehension of the Words He Speaks

https://youtu.be/kttVCbTrDLw

This brings me to the management of presentation. The biggest error people make when it comes to their presentation is when they focus exclusively upon the static components of visual cues. Even in the world of photo modeling where images are, by definition, static, photographers have developed the art of directing their models to emote and feel while they shift between various poses. It is the camera that turns the fluid poses of the model into static images. Therefore, the model's static images we see in fashion magazines are illusions. However, there are notable exceptions when it comes to the influence of static images.

Perhaps the example readers may recognize is the case of Martin Shkreli. Mr. Shkreli possesses a set of static facial visual stimuli that convey any number of negative attitudes, e.g., arrogance, insolence, disrespectfulness, pomposity, and being flippant, among many others. Add to Mr. Shkreli's static configuration a normal set of dismissive and condescending non-verbal cues, and the combination of the two (static and kinetic) create a synergy of numerous hyper-negative visual stimuli.

In 2014, New York Magazine's "Daily Intelligencer" section highlighted Mr. Shkreli's set of facial visual cues in a particularly vitriolic editorial entitled *The 10 Most Punchable Faces of Martin Shkreli.* [29]

Intelligencer / JUST WANNA SMACK THAT GUY

The 10 Most Punchable Faces of Martin Shkreli

By **Christopher Bonanos**

Martin Shkreli, hedge funder and pharmaceuticals profiteer, invoked his Fifth Amendment right not to testify on Thursday before the House Committee on Oversight. He also made a lot of smug faces, every one of which we would all like to punch.

Sarcastic eagerness.

Feigned interest.

Feigned benign pleasantness.

Seething sociopathy.

Basic smugness.

Sarcastic surprise.

Given what we know about how facial configurations modify the emotive centers of our CNS, Mr. Shkreli may arguably be the product of not only an unfortunate set of psychological attitudes, but also, he may be the victim of something he had little control over, i.e., his genotype-defined facial configuration (phenotype). At rest, Mr.

---

[29] New York Magazine (2014). Intelligencer: The 10 most punchable faces of Martin Shkreli, by: Christopher Bonanos.

Shkreli's facial configuration connotes arrogance and dismissiveness. His eyes and lip area are configured in such a way that whether Mr. Shkreli is actually feeling dismissive or snide, *he appears to be* conveying these negative attitudes. Human beings are not the only creatures that either benefit from or are negatively stereotyped by their natural facial configurations.

One creature that has benefited from its facial configuration is the dolphin. The dolphin's maxillofacial configuration looks like a friendly smile, to wit:

In fact, the dolphin's natural maxillofacial structure has spawned any number of **unwarranted projections** upon the dolphin, including human-like intelligence and healing powers, e.g., some women actually choose to give birth in the ocean with dolphins. [30]

It is the author's opinion that virtually all of these positive dolphin characteristics are anthropomorphized projections that are based, in large measure, upon the dolphins' naturally occurring facial structures and a few of their ocean antics, and have absolutely nothing to do with the natural behavioral characteristics of dolphins themselves. To learn more about the natural behavioral characteristics of dolphins, we refer the reader to the work of Barrett and Würsig. [31] It will come as a surprise to most readers that dolphins are not as sophisticated as chickens in

---

[30] Time Magazine (2013). *Dolphin-Assisted Births Are a Thing*. By: Melissa Locker.
[31] Barrett, L., & Würsig, B. (2014). Why dolphins are not aquatic apes. Animal Behavior and Cognition, 1, 1–18.

many ways, and often act like thugs. In fact, the UK Daily Mail summarized a six-year long study done at St. Andrew University in Australia with the following headline:

> **Flipper is a thug! Scientists reveal that dolphins are NOT as clever as other animals and are more likely to fight with one another**
>
> *Decades of scientific research suggesting dolphins have human-like qualities are flawed, according to new findings. It has been claimed that dolphins are less sophisticated than chickens.* [32] [33]

One final point about static visual cues: when it comes to static images, the vast majority of people find perfectly still images disturbing. In other words, all of us move all of the time, and when we don't move, we scare people. This fact is the primary reason that many people find dolls to be frightening and uneasy to be around.

Since we've already documented that our nonverbal behavior may be synergistic with our static image or may actually nullify, enhance or contradict it, people who aspire to be attractive cannot simply focus upon their static beauty. But that is exactly what the vast majority of people do—unless, of course, they are professional actors.

Nonverbal behavior necessitates that the presenter possesses the ability to feel positive emotions, and then display those feelings on the various channels of verbal and non-verbal communication. In this respect, beautiful nonverbal behavior comes from within. Nonverbal beauty, unlike static beauty, truly is captured by the old saw, "Pretty is as pretty does."

In today's politically correct world, people have been conditioned to not express their true feelings, lest they offend someone. Hence the promotion of the disingenuous smile, phony warmth (have a nice day) and fake static images that characterize modern social interaction dynamics.

---

[32] UK DailyMail (2013). Link: http://www.dailymail.co.uk/sciencetech/article-2415575/Scientists-reveal-dolphins-NOT-intelligent-likely-fight-other.html#ixzz4nyoBJ7WM
[33] Gregg, Justin (2013). Are Dolphins Really Smart? The mammal behind the myth. Oxford University Press, Oxford, U.K.

Clinicians like myself believe that one of the primary reasons drug abuse is rampant among performers is that, as a group, they are forbidden to experience, then display, the ups and downs of normal life. Performers must be "on" regardless of how they feel inside (The show must go on). And to make sure they are always "on," they often resort to using pharmaceutical assistance to ensure their "on" button remains "on" regardless of how they feel.

Central nervous system stimulants are the favored "on" drug among performers, which means that to turn the pharmaceutically induced "on" button "off," they must resort to either a sedative hypnotic or an opiate class of "off" drugs. Once the on-off cycle is regulated by pharmaceuticals, it is very difficult to stop one's reliance upon these drugs. Does any of this sound familiar when it comes to drug abuse in show business?

All of these issues are the direct consequences of being in a false presentation profession. Is it any wonder that beauty and acting are intertwined in a complex battle with one another? In today's social media intensive environment, we are only as good as our last performance.

> *Socially conformed presentation dynamics create people who are disconnected from their true selves, who then interact with other false selves in a poorly directed stage play comprised of disingenuous non-verbal visual and auditory cues, fabricated and politically correct content that we call life in the 21st Century.*

LIPS

Did you ever wonder what it is about lips that makes women want to make them bigger, color them, highlight them, and purse them when they flirt? For that matter, have you ever wondered why men seem to go crazy over a pretty pair of big red lips? What would you think if I were to tell you that a woman's lips are psychologically interchangeable with her vaginal lips? What did you just say, Doctor Napoleon? I know you heard me.

A key reason women with bigger and redder lips are more attractive is because fuller and redder lips are similar to vaginal labia. This connection occurs on the unconscious level in men, which makes the association even more powerful. But the lips/vagina connection doesn't end there. In fact, studies have shown that the larger a certain part of the upper lip is, the greater the likelihood that the woman with that particular body part configuration will be able to more easily experience orgasms. I said "**will be able to more easily experience.**" I did not say, will be able to **fake** orgasms.

That part of the upper lip is named, as irony would have it, the *labial* tubercle. This is the part of the upper lip in the middle that protrudes. It seems that the more prominent the labial tubercle, the more sexually responsive its owner. [34]

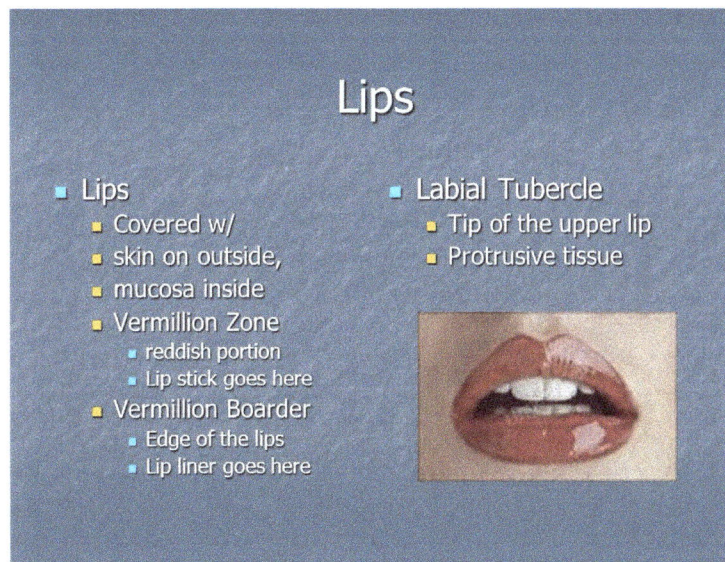

Studies have shown that waitresses who color their lips with a bright red lipstick receive larger tips. Makes you reconsider the entire issue of kissing, now, doesn't it? Actually, speaking of kissing (osculation) it seems that kissing has a distinct sociobiological function other than, as we have now learned, a faux oral sex act. Admit it, dear reader, you really did believe that a book on visual image would be shallow, right? I recall one hapless soul who remarked, "a book on that subject has to be dry."

---

[34] Brody S, & Costa RM (2011). Vaginal orgasm is more prevalent among women with a prominent tubercle of the upper lip. The journal of sexual medicine, 8 (10), 2793-9 PMID: 21676178

Kissable female lips serve a critically important sociobiological purpose. Kissable lips draw the male of the species' lips close enough so that the woman's limbic system can get a whiff of her potential mate's pheromones. And what, pray tell, do pheromones have to do with anything?

Well, it seems that women prefer men with an immune system different than their own, and women's limbic systems can sense these things when they kiss. Now short of having the man undergo a series of DNA blood tests, with particular reference to his HLA profile, the next best way to judge his immune system is to kiss him. When people talk about having chemistry with someone, I bet you didn't know that they were literally talking about organic chemistry!

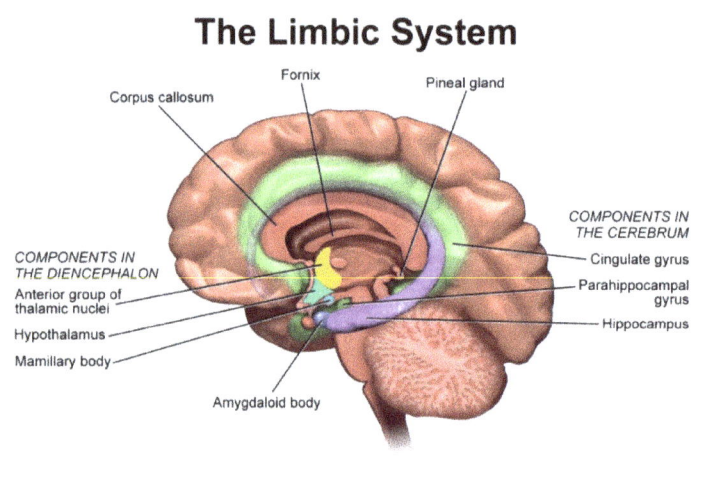

The Limbic System

Lips are unique among all facial features, because they don't sweat and they are as unique as a fingerprint. Lips are very sensitive and represent the first part of our body as infants that we love to stimulate. Yes, there really is such a thing as an oral fixation. The vast majority of us had it, and many of us still have it. Whether it was mother's breast, a baby bottle, a pacifier, your own thumb, a cigarette or cigar or vaping instrument, the reason that oral stimulation is so gratifying is that it stimulates the lips.

Why are lips so sensitive, you may ask? If you look at the areas of the human brain that are connected to certain parts of our bodies, you'll notice something very interesting. Some parts of our bodies are disproportionately represented on our brains' sensory topography. In other words, certain parts of our bodies take up a lot of real estate in our brain. Scientists have named the graphic drawing that illustrates our brains' sensory topography the "Homunculus."

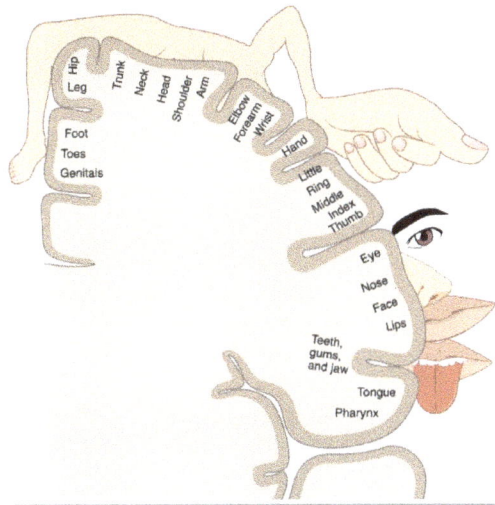

Homunculi

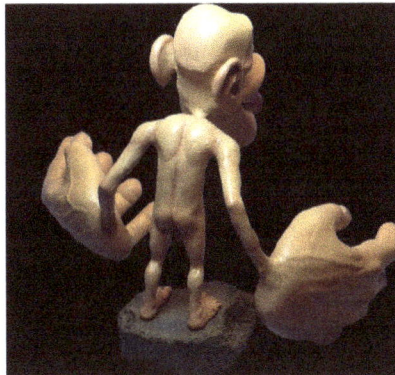

If you were to ask your brain what the biggest and most sensitive parts of your body are, your brain would name three areas:

1. Lips,
2. Genitalia, and
3. Index finger.

Oh, and speaking of index fingers. The same Scottish psychologist who documented the link between orgasms and labial tubercles, Dr. Stuart Brody, also found that the more sensitive a woman's index finger is, the more sexually responsive she is. Admit it, you THOUGHT that "psychology" was nothing more than what you see on your television when one guy or gal gives "advice" or "coaching" that sounds very similar to what your busybody aunty used to tell you.

Now, only my readers know (along with a few select researchers) why one of the most common nonverbal SEXUAL RECEPTIVITY behaviors on earth is the index finger rubbing across or simply touching our own lips. When you stimulate the lips, you are tickling your brain. Your brain loves to be tickled.

Two Attractive Women Tickling Their Limbic Systems (Dear Male Readers: hopefully while they are looking at you)

You see people tickling their brains all the time. For example, the oral gratification associated with smoking or vaping is inextricably linked to our lips' sensory topography on our brains.

If you watch one-year-olds, you'll notice that they experience the world through their lips and mouths. Young children go through a stage (aptly named the "oral stage") where anything they are interested in is picked up and either rubbed against their lips, or the objects they find to be most interesting, they insert into their mouths. What parent does not remember saying multiple times: "Don't put that in your mouth!"

Children going through their oral stage will fix their gaze upon mother's lips as though there is some magical property about lips that make them beautiful to gaze upon. One could reasonably conclude that young humans pass through a lip fetish stage. As we age, we pass through that phase but never quite leave our affection for lips totally behind. Those of us in the West who are fond of lips are not alone in the world.

As if to prove that man's love of beautiful lips is wired into our brains, archaeologists and anthropologists tell us that lip plates have independently developed as a popular body modification at least six times in known history. Lips were stretched so that they would fit lip plates going all the way back to 8700 B.C. among the tribes living in the Sudan and Ethiopia. Mesoamericans stretched their lips as long ago as 1500 B.C. Coastal Ecuadorians were modifying lips in a similar fashion around 500 B.C. Modern-day Africans, including the Sara and Lobi of Chad, sometimes remove the two upper and lower front teeth in order to fit lip plates into both the upper and lower lips. The Makonde of Tanzania and Mozambique put plates only in their upper lips.

Lip modification is inextricably linked to sociobiology. For instance, the Mursi tribesmen of Ethiopia begin the process of lip modification around 6 to 12 months before marriage. Researchers tell us **the size of the modified lips increases the economic value of the Mursi women.**

Mursi Tribeswoman

Understood from this anthropological/archaeological perspective, modern lip enhancement surgery and injections strike me as merely the latest iteration of human beings' desire to increase their sexual attractiveness—that is, their beauty and its economic value—by increasing the size of their lips either through surgery, piercing, makeup, or merely pursing one's lips as if you were kissing.

THE QUEST FOR BEAUTY

A huge part of the West's economy is rooted in our quest for beauty. Clothing, makeup, self-care products, and aesthetic treatments of all stripes and colors, including weight loss regimens, gyms, cosmetic surgery, beauty shops and parlors, exist for one reason: to satisfy our quest for beauty.

An even larger part of the West's economy is linked to the indirect quest for what beauty promises. As I documented in my essay on automobiles, a huge part of the marketing of cars is predicated upon how certain vehicles make their owners more attractive.

Should the economy of the West extract those parts involved with the quest for and maintenance of beauty, the West would immediately fall into a deep economic depression. And speaking of economic recessions and depressions, beauty-related products and services appear to be virtually immune to economic downturns. Look at the 2015 data related to the economy of beauty-related products and services:

"In 2015 the industry generated $56.2 billion in the United States. Hair care is the largest segment with 86,000 locations. Skin care is a close second and growing fast, expected to have revenue of almost $11 billion by 2018. This growth is being driven in part by a generally increasing awareness of the importance of skin care, but also specifically due to an increase in the market for men." [35]

US Beauty Industry Segments	Market Share by Revenue
Hair care	24%
Skin care	23.7%
Cosmetics	14.6%
Perfumes and colognes	9.5%
Deodorants, antiperspirant, feminine cleaning	8.5%
Oral hygiene	5.6%
Other	14.1%

When it comes to the men, the quest for power is often a sublimation of their frustrated quest to be attractive, as evidenced by how much money is spent on image-related products and services. What this means is that unattractive men often pursue great wealth in order to obtain the popularity that was denied them as young men by virtue of their looks.

The rich man/pretty woman matchup is but one manifestation of this sublimation phenomenon. The fact that beautiful women are attracted to rich men is why, in large measure, unattractive little boys aspire to be rich. Beautiful little girls begin life attracted to the attractive little boy. But when beautiful girls become

[35] Franchise help (2017), Beauty Industry Analysis 2017 - Cost & Trends.

women, many of them become dissatisfied with attractive men UNLESS those attractive men are also financially successful.

The sociobiological-rooted feelings and impressions at work are related to the fact that material wealth often follows success. Success in a competitive environment is a much better measure of reproductive viability than is merely winning the attractiveness lottery. And as if to create even more dissonance for women attracted to men who won the beauty lottery, versus those men who are rich, wealth and physical attractiveness are most often strange bedfellows.

As with any complex and nuanced subject, the details of that subject are where truth resides. For example, an attractive male actor can become rich. However, and I state this as an empirical fact, without any pleasure and with a great deal of compassion and empathy for actors: other actresses and actors do not typically view such a man as a viable reproductive choice.

This may be due to the fact that, in general, thespians do not make good long-term wives or husbands. Marriages often fail, infidelity is rampant in Hollywood, and lascivious behavior is virtually the norm. It may come as a surprise to the reader that many successful actors do NOT respect their craft. Some people believe that this is why, at least in part, many successful actors transition to the role of director and producer, two roles actors respect more than they do their craft.

Why so many attractive men fail to achieve career/financial success is a fascinating subject to ponder. On the one hand, male beauty provides its *youthful* owner any number of social advantages early in life. But as men age, it appears as though pure attractiveness and success become estranged from one another.

It appears that MERELY physically attractive men do not develop the life skills necessary to become successful and/or rich. (We rule out, of course, those few men who parlayed their looks into a well-paying career.) In large measure, this may be due to a dampened ambition to succeed in life, the direct consequence of beginning life with any number of freely given social advantages that simply do not continue into adulthood.

I've made scientific assessments of the world's richest men and their visual presentation. I've demonstrated the existence of a replicable negative correlation between wealth and male beauty. Not only were rich men (outside of image and entertainment businesses) not attractive, they were less attractive than the average man. Still, their girlfriends and wives, especially their girlfriends, were significantly more attractive than average-looking women.

To this day, one of the best predictors of the attractiveness level of a man's girlfriend and wife (girlfriend more so than his wife) is his financial success. Handsome *young* men tend to get all the *young* girls' attention and tend to be very popular during their *youth* by virtue of nothing more than winning the looks lottery. However, at some point in the male and female maturation process, women tire of and sometimes come to dislike the pretty boy in favor of men who have won the natural selection contest as manifest by the fact that he has achieved some degree of material success.

By the time attractive young men figure out the sociobiological rules—that is, if they ever do—it is too late for them to develop the character and skill sets necessary to prevail in the natural selection contest that is life. Thus, the pretty little boy is most often doomed to always play that role, even as he ages. And who, pray tell, will pair up with the pretty little boy turned adult man? The answer is a masculine protest woman who wants him as her possession and whipping boy.

The quest for beauty has many ironic aspects to it. One of the more dramatic ironies is the fact that beauty is a diminishing asset. This means that even those people who have achieved beauty have a lot in common with those who strive to be beautiful. In almost all human endeavors, winning the race is a time for celebration and relaxation. Consider for a moment: once you win the Indy 500, your timeless face will forever be enshrined on the Borg Warner Trophy.

Takuma Sato, Winner of the 2017 Indianapolis 500

A.J. Foyt, Winner of the 1977 Indianapolis 500

Indianapolis 500 Borg Warner Trophy

Unlike winning the Indy 500 race, winning the beauty race is a transient win. As soon as one crosses the beauty finish line, you begin to lose that which you just won. When women are younger, diminishing beauty goes unnoticed because it tends to drip away a drop at a time. One predictable phenomenon I have noticed is that while it is true that beauty drips away in imperceptive drops, the realization of the loss of beauty is often instantaneous. One day a woman awakens to look in the mirror and, as if overnight, she realizes that her beauty has vanished right before her very eyes.

Forgive the scientist in me, but women's all-at-once realization that they are no longer young and beautiful reminds me of a super-saturation experiment. In this experiment, crystals of salt can be dropped into water a crystal at a time, with no perceptible change in the water. One day, however, after thousands of salt crystals have been successfully mixed into the water, just one more small crystal of salt does something remarkable to the water. Want to see what happens? Remember, I'm showing this super-saturation experiment to you because this is often what is like for a woman who one day realizes her youthful beauty has vanished:

*Super-Saturation
Experiment Metaphor*

https://youtu.be/1y3bKIOkcmk

Once beauty's diminishing quality makes it into our consciousness, panic often sets in, and all of the warnings and platitudes about focusing upon one's inner self hit the image-conscious person like a ton of bricks. Those who have the financial wherewithal often seek out plastic, cosmetic or dermatological medical care.

The doctor's office becomes their temple, where the promise of salvation appears like a lighthouse to a ship's captain on a stormy night.   Alas, relief is only temporary until beauty is redefined as relative or loses all of its value, in the same way a fortune once had, but lost, forces one to embrace the virtues of poverty. Remember:

> *Beauty is serious business, more so for those who once had great beauty than for those denied its pleasures—and pain.*

GENOTYPE/PHENOTYPE

In previous chapters herein, I have made reference to the terms "genotype" and "phenotype." Now I want to explore these very important terms in more detail. As you will come to see, visual image is much more complex than you ever imagined.

Our visual image; that is, phenotype, is a reflection of our genetics; that is, genotype. Our phenotype can reflect an underlying unique chromosomal set; we label the configuration set of our chromosomes a karyotype.

At first blush, the reader may think that the ability to decipher a person's phenotype in order to ascertain a person's geno/karyotype is limited to geneticists or experts like myself. And while that is true when it comes to in-depth and complex analysis, I can teach people how to read *some* of the more obvious phenotypic clues that give away a person's geno/karyotypes. We've already learned one phenotypic expression that tells us about a woman's genotypic predisposition to experience an orgasm. Remember what that phenotypic visual cue is? The answer is the *labial tubercle*.

Because it is more illustrative and eminently easier to focus upon a person's karyotype, I'll provide the reader a few examples of how phenotypic configurations can be interpreted to reveal karyotypes.

Human beings nominally possess 23 pairs of chromosomes, i.e., 46 in total. Sometimes, however, an error occurs in mitosis, replication and/or rearrangements or assortments of chromosomal material. One of the errors than can occur involves the attachment of a third chromosome to the 21st pair of a person's 23 chromosomes. If a third chromosome attaches itself to pair 21, we call that "Trisomy 21." This particular karyotype is also known as Down's Syndrome.

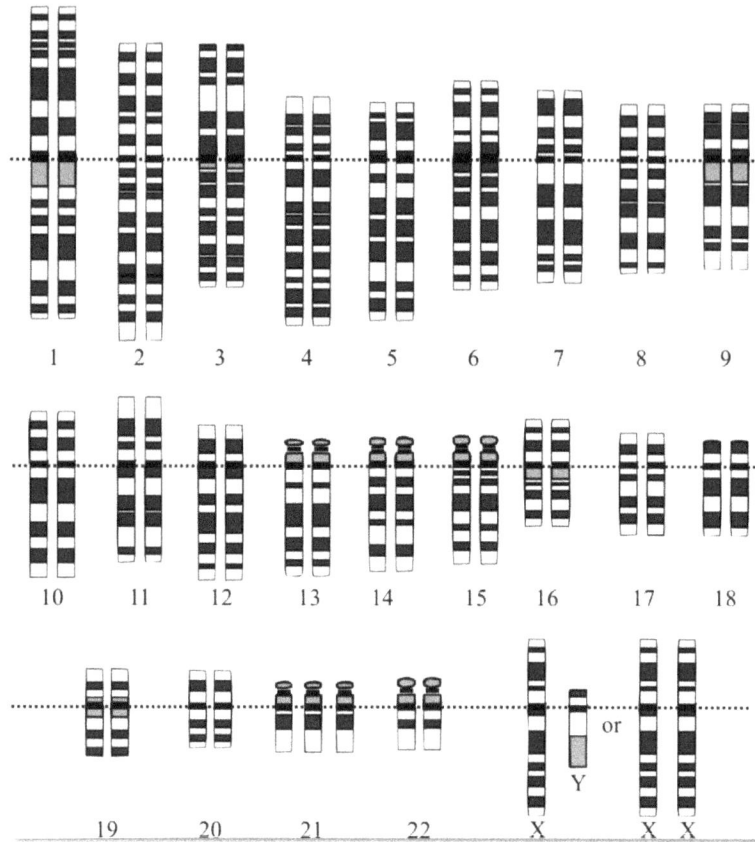

Trisomy 21 or Down's Syndrome Karyotype

What is interesting about people with the Trisomy 21 karyotype, as it relates to visual image, is that the Trisomy 21 phenotype is easily recognizable. In other words, the average person can learn to pick out the one person in a lineup of 50 people who possesses a third chromosome on the 21st pair of chromosomes with just a little training.

Phenotypes Characteristic of the Trisomy 21 Karyotype

Another karyotype anomaly that manifests in phenotype is called Klinefelter's Syndrome. Before I discuss this karyotype, a very brief discussion of the difference between autosomes and sex chromosomes is in order.

As we have already taken note of, the nominal configuration of a human's karyotype consists of 23 pairs of chromosomes;, that is, 46 chromosomes. 22 pairs are termed autosomes, and one pair is labeled sex chromosomes. They are called sex chromosomes because these two chromosomes determine sex or gender. Males have an Xy configuration (the y is denoted in lower case because it possess less DNA material) and females have an XX configuration.

In Klinefelter's Syndrome, the sex chromosome denoting a male (Xy) possesses an extra X chromosome; that is, the Klinefelter's karyotypic configuration is XXy, for a total of 47 instead of 46 total chromosomes.

Although Klinefelter's people are not as easily picked out from a lineup as are Trisomy 21 karyotypes, a person can learn to reliably pick out the male with the extra X sex chromosome using visual image cues. Males who have Klinefelter's Syndrome are in part taller, less muscular, have wider hips and larger breasts.

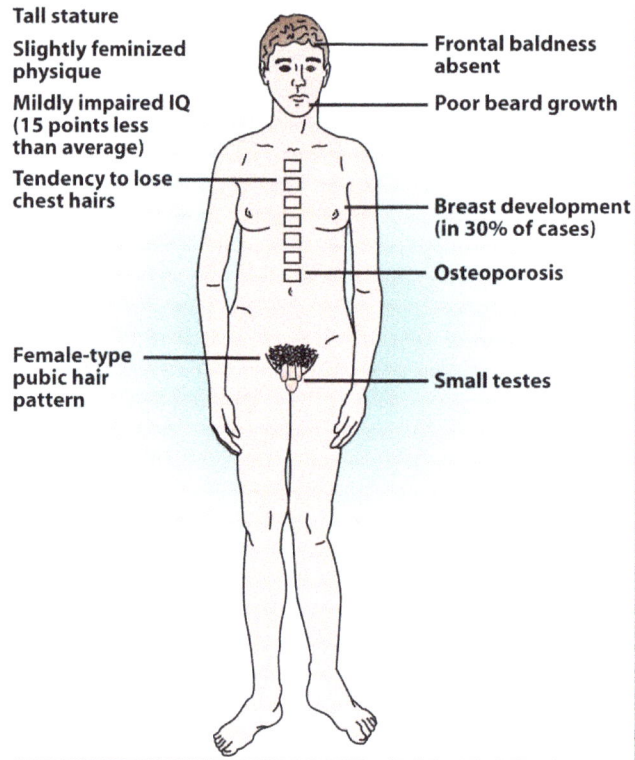

Tall stature

Slightly feminized physique

Mildly impaired IQ (15 points less than average)

Tendency to lose chest hairs

Female-type pubic hair pattern

Frontal baldness absent

Poor beard growth

Breast development (in 30% of cases)

Osteoporosis

Small testes

Phenotype Characteristics in Klinefelter's Syndrome

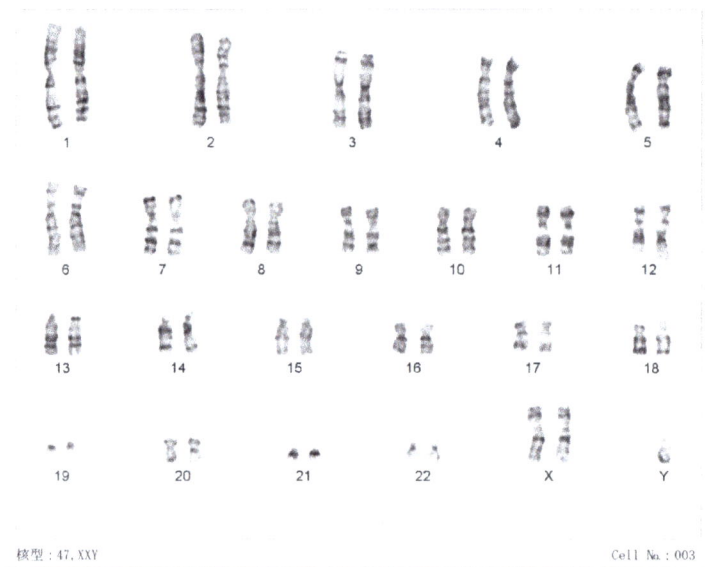

核型 : 47, XXY   Cell No : 003

Klinefelter's Karyotype

If Klinefelter's Syndrome people have one extra X sex chromosome, is there a chromosomal configuration where a female has only one X sex chromosome instead of the normal two X chromosomes? The answer is yes, and we call that Turner's syndrome; and like Trisomy 21 karyotypes, non-scientists with a modicum of training can spot the missing X chromosome simply by looking at women with this particular karyotype.

Turner's Syndrome girls are shorter and have poorly developed secondary sexual characteristics. Turner's Syndrome girls are stocky, have arms that turn out slightly at the elbow, possess a receding lower jaw, a short webbed neck, and a low hairline at the back of the neck.

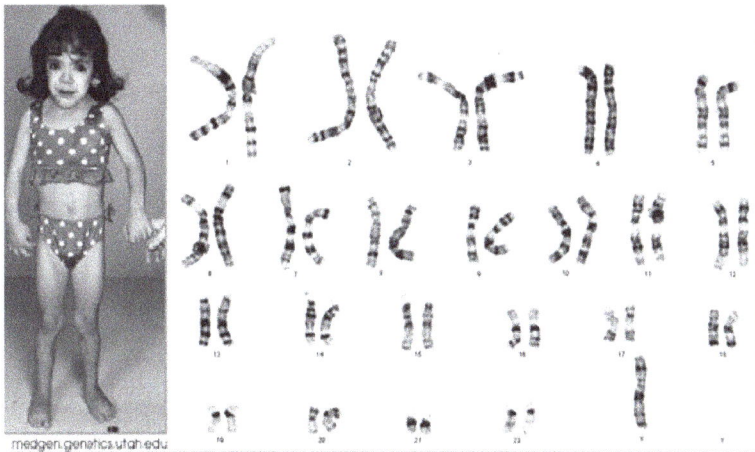

Turner's Syndrome Karyotype and Representative Image (Medgen Genetics)

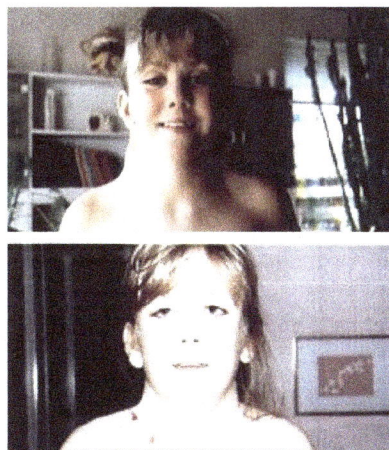

Turner's Syndrome Phenotype Showing Pre-Post Surgery for Webbed Neck

Notice that in the last image shown, plastic surgery was utilized to modify a phenotypic anatomical manifestation of an underlying karyotype. The important point I want to make here is that although the underlying karyotype of this patient was NOT changed, the phenotype was changed.

Cri du Chat is French for "cry of the cat." It is also the name of a genetic condition where the fifth pair of the normal 23 pairs of human chromosomes has one chromosome that is missing genetic material. We refer to the Cri du Chat karyotype as "a deletion on the short arm of chromosome 5." As such, it is also referred to as a 5P-Deletion karyotype.

As in our other genotypic/karyotypic anomalies, the 5P-Deletion is particularly interesting because not only is it visible, but also, as the name implies, you can hear it. Infants with the Cri du Chat karyotype have a modified voice box that has the effect of making the child's cry sound similar to a cat's cry.

The Sound of a Child
with Cri du Chat
karyotype

https://youtu.be/TYQrzFABQHQ

In addition to their unique voice box characteristics, these people are recognizable because of their small heads (microcephaly), an unusually round face, small chin, widely set eyes, folds of skin over their eyes, and a nose with a characteristically small bridge.

Cri du Chat

- 5p- syndrome
- Deletion of part of the 5th chromosome.
- Intellectual disability, delayed development, small head size, low birth weight, low muscle tone.
- Infants have a cry that sounds like a distressed cat.

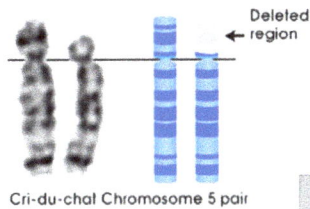

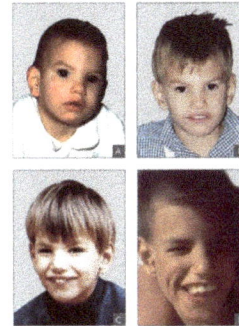

Cri-du-chat Chromosome 5 pair

Cri du Chat Karyotype and Representative Images

"80 percent of the 5P-Deletion cases, the chromosome carrying the deletion comes from the father's sperm rather than the mother's egg. Deletions during the formation of an egg or sperm are caused by unequal recombination during meiosis. Recombination normally occurs between pairs of chromosomes while they are lined up at the metaphase plate. If the pairs of chromosomes don't line up correctly, or if the chromosome breaks aren't repaired properly, the chromosomes can gain or lose pieces. Unequal recombination at a certain location on chromosome 5 causes cri-du-chat syndrome." [36]

What if I told you that sexual orientation can be determined with a fair degree of accuracy by utilizing software that analyzes various phenotypic cues? Stanford University researchers have created software that can distinguish homosexual from heterosexual faces. The software, known as VGG-Face, converts photographic images into a series of numbers. An algorithm compared facial features to predefined parameters to determine if the people in the photograph were homosexual or heterosexual. The software proved to be 81 percent accurate when it came to analyzing a single photograph of each of the male subjects, and 74 percent accurate when it came to judging a single photograph of each of the female subjects.

---

[36] University of Utah, Genetics Learning Center (2013) Cri du Chat genetics.

When the dataset was increased to five images per subject, the software's accuracy increased to 91 percent for men and 83 percent for women. So what are the sexual orientation visual image cues that manifest in a person's phenotype?

The authors stated that a "gender-atypical facial morphology" was associated with homosexual men. This included such cultural-based visual cues as grooming styles and more sociobiological-based visual cues such as facial expressions, narrower jaws, longer noses and larger foreheads. Lesbians were found to have smaller foreheads and larger jaws than heterosexual women.

The sociobiological visual cues related to facial morphology are consistent with parental hormone theory (PHT) of sexual orientation. PHT targets intrauterine environments, e.g., the exposure of fetuses to androgens. [37] In previous writings, the current author has discussed the impact of the intrauterine environment of the developing human being, to wit:

> *"Once the sex identity blueprint is laid down, the developing child spends the first 9 months of his or her life in mother's womb. This maternal environment is defined by the biochemical characteristics of the mother, herself. A maternal environment that is neutral, that is, neither flooded with male or female hormones, doesn't modify or alter the development of the child's genetically programmed sexual and future gender identity. A maternal environment that includes an excess of male hormones, e.g., androgen, as would occur in some women whose adrenal glands are hyperactive, can and often do accentuate the Xy developing child to make him "hyper-male" and make the developing XX child to become more male like.*
>
> *Any number of psychiatric conditions have been shown to be influenced by hormonal modification to the intrauterine environment of the mother. Anorexia Nervosa is one such condition. Doctors Procopio and Marriott published research in the Archives of General Psychiatry in*

---

[37] Kosinski, Michael, Wang, Yilun. (2017). Deep Neural Networks Are More Accurate Than Humans at Detecting Sexual Orientation From Facial Images. Journal of Personality and Social Psychology (in press), September 7, 2017.

December of 2007 (12):1402-7, that demonstrated changes to mother's intrauterine hormonal environment increased the risk of developing anorexia nervosa in female children, but not in male children.

Lesbianism has been positively associated with intrauterine environments that are androgen rich. Here is what authors Dreger A, Feder EK, Tamar-Mattis A. wrote in their study published in the journal of Bioethics, 2012 Sep;9(3):277-294.

'Following extensive examination of published and unpublished materials, we provide a history of the use of dexamethasone in pregnant women at risk of carrying a female fetus affected by congenital adrenal hyperplasia (CAH). This intervention has been aimed at preventing development of ambiguous genitalia, the urogenital sinus, tomboyism, and lesbianism.'

As the article noted, some parents, upon discovering that the pregnant mother has an androgen rich intrauterine environment secondary to CAH, have chosen to be treated with a substance named dexamethasone. Dexamethasone blocks the effect of the excess androgens circulating in the mother's blood. This is done in order to reduce the chances that their XX developing child will become a lesbian, develop ambiguous external genitalia, as well as any number of other physiological and behavioral characteristics. "[38]

I included this section on genotypes and their phenotypic manifestations because I want the reader to think about the fact that educated people can actually SEE what occurs at an invisible molecular level in the human body, IF you know what you are looking for and you know what you are looking at. Now let's apply this point to beauty.

---

[38] Napoleon, Anthony (2015). Transjenner: The Myths, The Manipulation and The Truth. https://gumroad.com/dranthonynapoleon

Beauty, like our aforementioned karyotypic anomalies, is a reflection of what is going on at the molecular level in our DNA. If the gravity of that fact escapes you, you've missed one of the most profound insights in all of sociobiology and, in particular, the research on visual image and beauty. Just ponder the significance of this fact:

An expert in visual image interpretation and genetics can see the footprints left in the snow of our DNA.

Phenotypic visual cues speak volumes about people. Keep in mind that your looks are very hard to hide, unless of course you wear a mask in public. By the way, experts such as myself are not hindered one iota by makeup or lighting. As you might imagine, we can easily redact, from our analyses, the contaminants of makeup and lighting, along with many other potential confounds. Nevertheless, the average person knows more about the genetic roots of visual images than he ever imagined.

Case in point: the average person can easily visualize genotypic racial differences. Virtually everyone can distinguish Sub-Saharan African genes from Northern European genes, Asian genes from Aboriginal genes, or Amerindian genes from Caucasian genes simply by looking at photographs of these various people. If race were merely a social construct, a person not exposed to and/or educated about the "racial social construct" would be unable to distinguish Sub-Saharan genotypes from Caucasian genotypes. It simply would be impossible.

Average people can, with 100 percent accuracy, distinguish "Little People" or dwarf genotypes from non-dwarf genotypes. And it is not just the visual cue of stature that serves as a phenotypic cue. Average people use visual configurations, proportional cues, color and its various qualities, e.g., wavelength and depth of color concentration, hue, etc.

Using Photoshop, it is easy to make equivalent the otherwise naturally occurring skin color differences between African and Caucasian people. Yet, despite making the respective groups' skin colors identical, the genotypic differences between African and Caucasian people, as manifest in their respective phenotypes, are readily identifiable. In fact, removing visual stimuli related to color does not affect the percentage of test subjects who can accurately identify Sub-Saharan African populations from Caucasian populations.

Albino genetic variations within the African genotype provide us confirmation of our assertions as stated above. Look at the following photograph of an African boy with Albinism:

An African Boy with Albinism with a Group of Peers

Also, simply by looking at phenotypic presentations, one can easily identify variations within a particular racial genotype. For example, the Pygmy genotype can easily be distinguished from their geographical neighbors, the Bantu. Height, a phenotypic manifestation of man's genotype, allows anyone with reasonably good vision to distinguish the fascinating sociobiological differences between the Bantu and Pygmy tribesmen of Cameroon. Included for the reader's consideration are some excerpts from a fascinating article that appeared in Live Science on this very subject:

> *"Why the Pygmies of West Africa have such short stature, while neighboring groups don't, has been somewhat of a mystery. Now new research suggests unique changes in the Pygmy's genome have both led to adaptations for living in the forest as well as kept them short.*

Researchers analyzed the genomes, the "building code" that directs how an organism is put together, of Western African Pygmies in Cameroon, whose men average 4 feet, 11 inches tall, and compared them with their neighboring relatives, the Bantus, who average 5 feet, 6 inches, to see whether these differences were genetic or a factor of their environment.

"There's been a long-standing debate about why Pygmies are so short and whether it is an adaptation to living in a tropical environment," study researcher Sarah Tishkoff of the University of Pennsylvania said in a statement. "Our findings are telling us that the genetic basis of complex traits like height may be very different in globally diverse populations."

[T] he researchers analyzed the genomes of 67 Pygmies and 58 Bantus for changes that would provide information about an individual's ancestry. These changes are small, non-harmful misspellings in the code (the chemical bases A, C, T and G) that makes up the genome. For example, a Bantu might have an A where a Pygmy has a T.

By analyzing large numbers of these changes, researchers can tell how much of an individual's genome is Bantu and how much is Pygmy.

Selected for stature

The researchers also used this letter-change data to look for areas of the genome associated with height and those that were "naturally selected" for — parts of the genome that are passed down through the generations because they provide some sort of survival advantage.

The data revealed height had a genetic component related to Bantu ancestry: The more Bantu ancestry an individual from the Pygmy tribe had, the taller that individual tended to be. One part of the genome, on chromosome 3, was especially important in this trait, the researchers said.

"We kept seeing a lot of them [these single-letter differences] highlight that region in chromosome 3," Tishkoff said. "It just seemed like a hot spot for selection and for very high differentiation and, as it turns out, very strong association with height as well."

Height genes

The researchers zoomed in on the genes in this area of the genome. One of the genes they found had already been associated with height changes in other populations, but the rest hadn't.

They found new changes in hormone pathways and immunity that seemed to correlate to the pygmy's short stature. These could have been selected for because of their influence on height or because changes in these genes play other roles in the body that were advantageous to the Pygmies, Tishkoff said.

For example: An immunity component might be selected for because it helps the pygmies fight off infections, which are prevalent in their habitat. And the link to hormone pathways also makes sense, Tishkoff said, because changes to them could help the Pygmies reproduce at earlier ages. Shorter height could just be a byproduct of these changes."[39]

---

[39] LiveScience (2012). Why Pygmies of Africa Are So Short. Article by: Jennifer Welsh, April 26.

Pygmies from Cameroon (Photo by Evan Leach)

A Group of Bantu Women

Let me repeat, I want the reader to appreciate that all of the genetic, sociocultural, anthropological, sociobiological and historical backdrop of Cameroon Pygmies and their neighbors, the Bantu, are reflected in visual cues. By the way, do Bantu and Pygmy populations, as a rule, find each other to be beautiful? No, they do not.

In consideration of the fact that genetics plays a pivotal part in reproductive selection and viability, is it any wonder that man evolved to be able to ascertain key facts about genetics by looking at people long before he was able to actually develop an understanding about how reproduction at the chromosomal level actually occurs?

This is why, in part, that constellation of visual image we call "beauty" is such a fascinating subject, because people everywhere rely upon beauty to become sexually aroused. Studies conducted at the National Institutes of Health (NIH) have demonstrated remarkable jumps in the amount of circulating testosterone in men who spent just a couple of hours looking at beautiful women dancing nude. Take those same men and have them spend two hours watching old and unattractive women dancing nude, and their testosterone levels will drop on average 33 percent. Women have a similar sexual response to handsome men. We choose our food, sexual partners, cars, homes, and virtually everything else based upon our wired-in love of beauty.

People denied and/or oblivious to the subjects of sociobiology and genetics, including the pathophysiology and inheritance coefficients of disease, are shocked to learn of the fact that our phenotypes reflect our genotypes. This is especially true for those brainwashed to believe the falsehood that such matters are social constructs.

A concerted effort has been made in America, beginning in the 1960s, to transform every sociobiological and genetic reality into a catalogue of social constructs. That level of brainwashing, which has been very successful, I might add, would be like redacting the genetics of eye color and explaining variations in eye color to the CHOICE of colored contact lens.

I can't take the time here to explain the who, what, when, where, why and how this social construct brainwashing was designed and carried out, but I will at this time provide the reader just one sliver of factual data that illustrates the utter nonsense of this particular social construct ideology.

Have you ever wondered why your physician typically asks you questions about your family's medical history, gender and race? It is because diseases have inheritance coefficients greater than zero and often significantly high, or is it because your physician is ignorant of the wisdom of social constructs or that he is a racist? A person inherits their genetic material from their family of origin (mother and father, etc.) AND their tribal family.

Take a look at the following group of people:

One of these men will have a Glomerular Filtration Rate (GFR)[40] that is so different when compared to the other three men, that a completely different scale must be used to calculate his GFR when analyzing a renal panel blood test.

If this person's nephrologist (kidney specialist) chose to treat this unidentified person pictured above as if he was no different from his other three colleagues, when it came to his GFR, that nephrologist would be guilty of medical malpractice. A GFR is NOT, I repeat NOT, a social construct.   GFR is significantly different between African originated genotypes and Caucasian originated genotypes.

Just think, no matter your level of education or political ideology, you are able to identify which person pictured above possesses a GFR so different from the other three men pictured above, that a different scale of what is and what is not normal is used by kidney specialists ALL OVER THE WORLD. Permit me to give you another example.

The photo that follows is of the Congressional Black Caucus (circa the Obama years). Two people among these men and women possess a genotype that puts them at a higher risk for skin cancer. Of those two men, one man has an exponentially higher risk of developing skin cancer. Using phenotype alone, can you identify those people?

---

[40] Glomerular filtration rate (GFR) describes the flow rate of filtered fluid through the kidney.

United States Congressional Black Caucus and Presidential and Clergy Guest

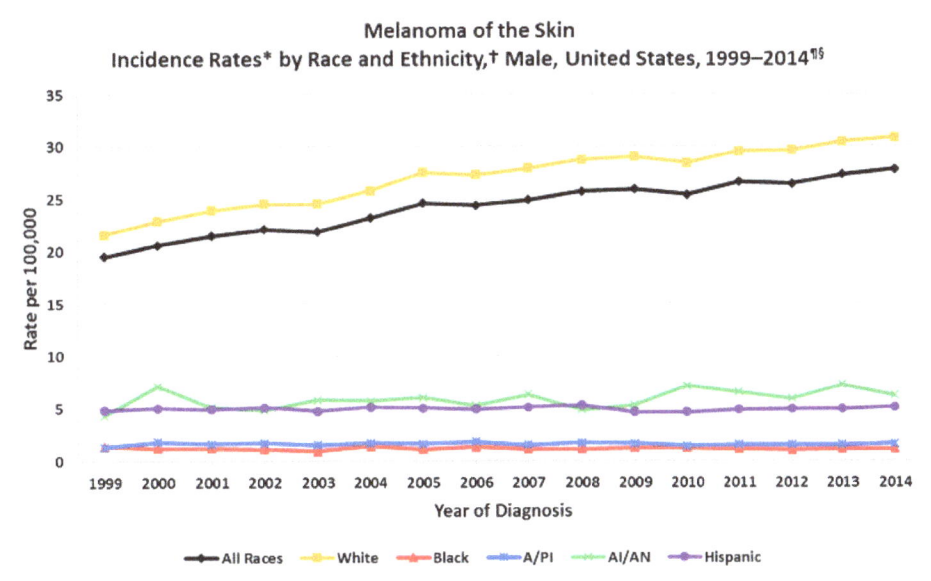

Centers for Disease Control and Prevention (2014 Data) Melanoma Skin Cancer Rates

You should easily be able to identify, using phenotype alone, which persons in the photo immediately above are at a greater risk to suffer from the most deadly form of skin cancer, i.e., melanoma.

Keep in mind that I have provided to the reader ONLY TWO examples from hundreds, if not thousands, that would serve to prove the same point. Phenotype can, to the ***educated*** eye, provide lifesaving and critically important clues about a person's genotype. Note that I highlighted the word "educated." The uneducated, the prejudiced or ideologically driven person who PRESUMES he or she can deduce genotypic predispositions from phenotype, is simply exercising his or her confirmation bias in order to promote their particular ideology.

Just as was noted when I discussed male and female mating behavior, those uneducated readers who are susceptible to confirmation bias have notoriously misinterpreted books on sociobiology to justify their mostly ignorant and often prejudiced ideologies. I repeat my admonition to the readers prone to confirmation bias. For every "defect" or "weakness" you may think you've learned from phenotype recognition of people not of your tribe, I can take YOUR reference group and show you a set of different, but equally "defective" markers that reflect YOUR genotype. Keep that in mind as you process the information herein.

When a certain genotype is less viable than a disparate tribal genotype, within any given geographical area, it is almost always because the geographical location of where that disadvantaged genotype evolved has been abandoned for an alien land, where the natural selection pressures that helped to construct that person's genotype are no longer salient.

For example, an indigenous Northern European person relocated to equatorial Africa is at an exponentially greater risk of developing skin cancer than the people whose genotypes are native to that land. Another example is sickle cell anemia. It is most common among people whose ancestors come from Africa; Mediterranean countries such as Greece, Turkey, and Italy; the Arabian Peninsula; India; and Spanish-speaking regions in South America, Central America, and parts of the Caribbean.

People with this disease have some immunity to malaria. Experts estimate that protection against overall mortality to be approximately 60 percent. So, people with sickle cell disease who live in areas where malaria is prevalent have an advantage over people without the disease. However, relocate these same people to an area like the United States, for example, where malaria has been all but eliminated, and their sickle cell disease puts them at a marked disadvantage.

RUTHLESS COMPETITION

Human beings manipulate how they present themselves to others, whether we are aware of this fact or not. Our manipulation of our visual cues, as reflected in our wardrobe, makeup and other behavioral choices, is to enhance our beauty by hiding faults, accenting sexual cues, and by displaying our *modified* viability visual cues. One distinction that I find useful in understanding our displays of beauty when it comes to wardrobe and accessories, including makeup, involves the differences between fashion and fads.

Fashion is timeless in the sense that it does not vacillate in response to the whims of trendy popularity. Skin tone defines fashion's color palette just as body type defines cut and style, both of which are elements of fashion and don't tend to change over time. Fads, on the other hand, are transient changes in wardrobe, makeup or related visual cues that can and often do trigger a mass adoption or trend. Fads come and go, while fashion tends to be more permanent.

Wardrobe and makeup fads serve as heightened sexual cues that are unconsciously designed to give men and women a temporary reproductive advantage. Therefore, when a woman sees a sexually alluring woman wearing a faddish shoe, skirt, pant or makeup style, she likely *feels* (whether her conscious mind picks this feeling up or not) a primitive drive to emulate what she sees, lest she be left out of her gender's sexual competition. As a result of this competitive sex drive, the wardrobe or makeup fad spreads like a virus until it becomes so widespread that it reaches critical mass, where almost all reproductive age women have adopted the fad. Once a critical mass number of people have adopted a fad, the fad loses its power, and it then fades.

We associate viral outbreaks of mass wardrobe or makeup adoptions with certain decades. For example, if I were to mention "flowers in your hair" you would probably know, if you are a student of mid-twentieth century history, that I am referencing the decade of the 1960s. If I were to mention bellbottom pants, you'd probably know I am talking about the 1970s.

Wardrobe or makeup fads are consciously recognized as "the latest style" or the "in-thing." People don't fully understand what it is that motivates them to copy a fad or feel like they are missing out if they don't adopt a trend. Sadly, the average person denies they are affected by trends, though replicable studies have shown, beyond a shadow of a doubt, that average people are trendy in every sense of the word. In reality, wardrobe and makeup fads are serious business. The motivation to adopt trendy visual cues in the form of fashion or makeup fads resides deep within our unconscious. It is the unconscious nature of trendy motivation that accounts for, in large measure, the average person's unawareness of what they do and why they do it.

The adoption of fads is a way for reproductive-aged men and women to display to others like them that "I'm in the game." My comments here are reserved for reproductive-aged humans, but realize that long after the reproductive years end, the psychological motivation to sexually compete continues.

In the animal world, we don't see wardrobe or makeup fads because other creatures don't have the option of adorning themselves with add-ons in the form of wardrobe, makeup or transportation choices. But if other creatures could accessorize themselves with add-ons, what you would see is exactly what you see displayed in human beings. It wouldn't take long before the peacock would change the shape of his feathers or their colors to try to gain a reproductive advantage over his fellow peacocks.

Wardrobe and makeup fads draw attention to the adopter of the fad, but not in the traditional sense. Women are exquisitely attuned to what other women are wearing or visually cuing that may put them at a sexual disadvantage. Sometimes just hearing about a particular wardrobe, perfume or makeup fad is enough to trigger women's sociobiologically driven competitive urges to adopt the fad. These drives manifest as the obsessive need to have what has "worked" for another woman.

What do I mean when I say "worked?" I mean reproductive success or, at least, *going through the motions* of reproductive success. For example, mating with a powerful man is an example of "working" from a sociobiological prism's filter.

Becoming the lustful object of rich and powerful men is another example of "working." Women who do become the sexual object of desire of powerful men motivate other women to emulate them, down to the tiniest detail. Let me give you a couple of classic examples from recent history.

When Monica Lewinsky had a sexual liaison with one of the most powerful men in the world, President Bill Clinton, she triggered many women's ruthless competitive urges to emulate her and/or not be outdone by her (another female's eggs). After all, Miss Lewinsky had managed to attract the sexual attention of a powerful man. What was it about Monica that gave her an edge? This was the question asked on the *unconscious level* by millions of women everywhere.

When it was revealed that Miss Lewinsky wore a particular red lipstick manufactured by Club Monaco, **millions of women** immediately went out and bought Club Monaco red lipstick. Because of the contagion effect, i.e., fad, the result of women's ruthless sexually competitive urges, Club Monaco could not keep their red lipstick on store shelves.

Perhaps the reader noticed that after Governor Sarah Palin gained notoriety as John McCain's relatively attractive vice presidential running mate in 2008, millions of women began to do something different; something they would have never done, had not this particular famous woman done it first. What was it that Governor Palin did? She wore glasses. Almost overnight, I saw women of all ages and across all demographic groups wearing glasses—and not just any glasses, but Palin glasses!

Sarah Palin's glasses boost frame's sales

L.A. JOHNSONPITTSBURGH POST-GAZETTE

Men tend to display their ruthless competitiveness differently than do women; nevertheless, the same dynamics are at play. Men tend to adopt certain body hair patterns and drive "manly" machines and adopt mindsets, as opposed to buying wardrobe, makeup or shoes, though men do that as well. Male peacocks love to show off their beautiful lady friends to other men. Men tend to adopt moustaches, sideburns, and longer or shorter hair as their sexual display fads. Once a popular-among-the-ladies male dons a moustache, for example, other men tend to adopt the facial hair pattern that "worked" for the "other guy."

Certain decades are associated with identifiable presentation styles for men. In the mid-1960s, The Beatles were lusted after by throngs of adoring teen girls. The Beatles' popularity with young girls did not go unnoticed by those "in the game" and singlehandedly created the fad of relatively longer hair for men or, as it was termed in the day, the mop-top fad.

The decades of the late 1970s and early '80s found men wearing longer sideburns and donning moustaches. 1980s television star Tom Selleck and his character Magnum P.I., a Ferrari-driving babe and gay man magnet, helped to make moustaches all the rage.

Madonna defined the layered and lace look of the 1980s, while the 1990s were all about bleached hair, one pant leg rolled up, bike shorts and bare midriffs. Britney Spears and Christina Aguilera made popular, i.e., faddish, the transition from innocent pre-pubescent teen to slutty teen wear in the brand new millennium.

The new millennium marked the coming of age of the generation known as Millennials. The latest generation, Gen Zs, have made presentation choices that reflect any number of profound cultural changes. For example, take Gen Z's underwear choices. Girls tend to wear almost exclusively thong underwear. Only fashion historians know that thong panties were once *only* reserved for strippers and ladies of the evening, as recently as the 1970s to early 80s.

And it is not just Millennials and Gen Z's underwear choices that have changed. Besides wearing thong panties, younger Millennials and older Z'ers have adorned themselves with tattoos, candy-colored hair, and body piercings such as nose rings, studs and belly button jewelry, just to name a few of the most popular body modifications. Why did these particular generations adopt such dramatic visual cues that scream, "I'm here, look at me?"

Millennial and Gen Z males have been scolded since they were little boys for displaying their masculinity. The reason for this, in large measure, has to do with the huge numbers of Millennial and Gen Z boys who were raised by single mothers; that is, without a father or strong male role model in the home.

Female- & male-headed families
(of families with children under 18)

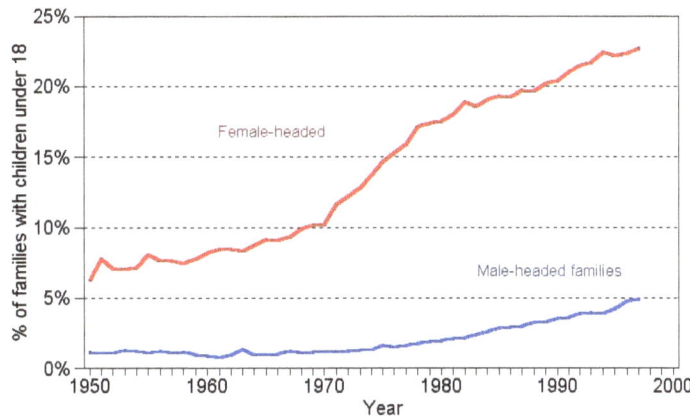

Source: www.census.gov/population/www/socdemo/hh-fam.html

Single mothers raised their Millennial and Gen Z sons as imperfect females. Any behavioral display of traditional masculinity was shamed, conditioned out or, in some cases, drugged away with psychiatric medications. By shunting male's displays of masculinity, the incidence of androgynous males skyrocketed. As a consequence of this male-gender trait tamping down:

> **Millennial and Gen Z girls became hyper-attention seeking in order to arouse the suppressed masculine sociobiological identities of their male counterparts.**

Millennial and Gen Z males have not only had their sexuality modified by genocidal cultural tsunamis, but biological changes have also been at work. For example, Millennials and Gen Z males have, on average, 20 percent less testosterone than the generation of males who came of age in the 1980s. This reduction in testosterone has manifest in wardrobe and other behavioral choices that scream the absence of testosterone and masculinity.

A Typical Group of Gen Z Boys and Girls

If the reduced amount of male hormone circulating through Millennials and Gen Z male's bloodstreams was not enough of a liability, the zeitgeist cultural norm that scolds boys for being boys has all but behaviorally neutered these generations' males.

Let's look at the sexual arousal issue from the Gen Z female's point of view. Imagine if your wardrobe and makeup, i.e., visual sexual cues, are tasked with arousing teen boys who have reduced amounts of testosterone and have been conditioned to shun male behavior? Not so long ago, and spanning from time immemorial to sometime in the 1990s, teen boys were at the height of sexual reactivity. But this is no longer true.

Is it any wonder that Gen Z females color their hair using a color palette right out of a Yu Yu Hakusho anime, affix shiny jewelry to every orifice that lends itself to such piercing, wear over-the-top fake eyelashes, and are sexually forward, if not aggressive? Being frustrated with an opposite sex that isn't all that opposite to you adds a degree of frustration to *la danse sexuelle* for Gen Z girls never experienced before by any other previous generation of females.

Gen Z young men wax their bodies, gel their hair, wear eyeliner, and as the 1970s song that ushered in the trend toward androgynous sexuality forewarned us, "Dude looks like a lady."

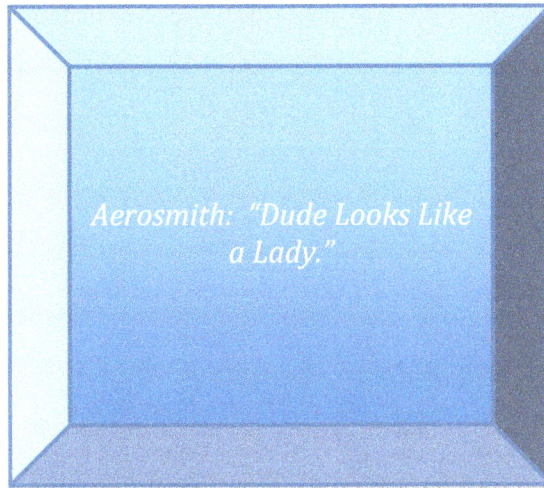

Aerosmith: "Dude Looks Like a Lady."

https://youtu.be/nf0oXY4nDxE

As I write this chapter in 2017, knee-high leather boots are all the rage among Millennial women and some Gen Z girls. Why do these particular generations adorn themselves in boots that resemble the footwear worn by Nazi SS circa 1943? What war are these women fighting? To answer that question, take a look at a few data samples that capture America's current gender zeitgeist:

Bachelor's Degree or More at Ages 22, 23 and 24, by Gender

Source: BLS

Age 22	Age 23	Age 24
Male 6.8% Female 12.7%	Male 14.3% Female 23.4%	Male 18.7% Female 27.6%
187 Women per 100 Men	164 Women per 100 Men	148 Women per 100 Men

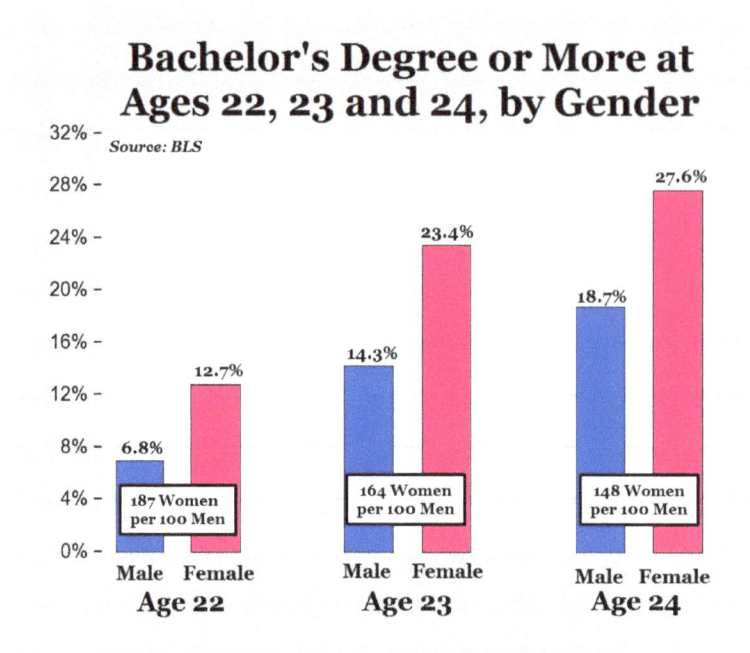

College Degree Gap Between Men and Women 2015

"In 2015—the year for which the most recent data are available—42 percent of mothers were sole or primary breadwinners, bringing in at least half of family earnings. Nearly another one-quarter of mothers— 22.4 percent—were co-breadwinners, bringing home from 25 percent to 49 percent of earnings for their families. This represents an increase over previous years and is the continuation of a long-running trend, as women's earnings and economic contributions to their families continue to grow in importance."[41]

---

[41] Center for American Progress, (2016), *Breadwinning Mothers Are Increasingly the U.S. Norm.* By: Sarah Jane Glynn.

FIGURE 3. Percentage of Men Never Married, by Age

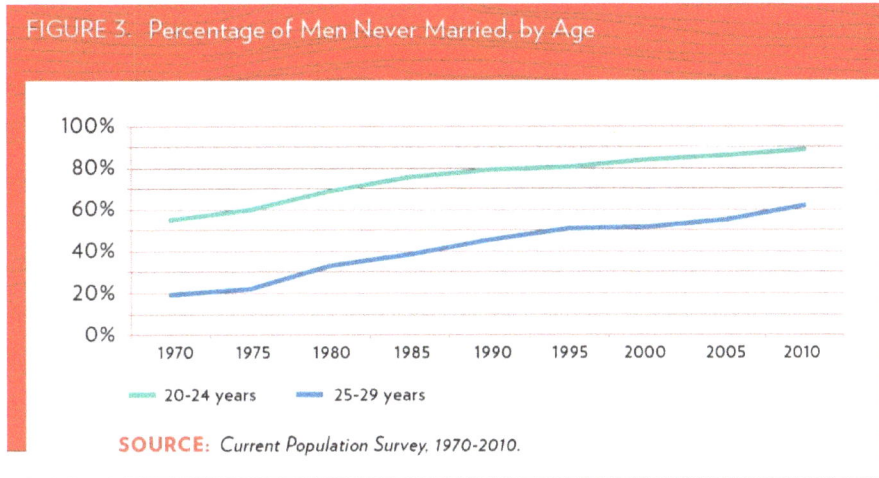

SOURCE: Current Population Survey, 1970-2010.

As these data point out, Millennial and Gen Z girls have or are about to realize they are on their own. My point in addressing these socio-cultural issues is this:

> *How we present ourselves as reflected in our wardrobe, jewelry,*
> *footwear, tattoos and body piercings reflect our cultural zeitgeist.*

What war are women fighting? It is an economic and reproductive war, born of a genocidal cultural revolution that has feminized men and left women to fend for themselves while desperately trying to hold onto some semblance of sexual allure for the benefit of *an increasingly unreceptive audience*. Is it any wonder that women's footwear reflects this reality?

A

Random Selection of Women's Boots for Sale Online 2017

The SS Look

The all-black tunic and peaked cap at left above were worn by a *Scharführer*, or staff sergeant, of the Allgemeine-SS Standarte 45, a regiment based at Oppeln in Upper Silesia. The red border on the cuff band indicates that the wearer belonged to the 3d Battalion. A braided aluminum chin strap marks the cap above as an officer's hat. The dress boots were worn by all ranks.

Nazi SS Wardrobe Ensemble

MISCELLANEOUS THOUGHTS ON BEAUTY

Physical Attractiveness and Mental Illness

The very first scientific study I conducted on visual image investigated the relationship between physical attractiveness and mental illness. A researcher by the name of Amerigo Farina, Ph.D., had conducted research at the University of New Mexico. Dr. Farnia found that people who suffer from mental illness are less attractive. Farina studied patients in a mental hospital.

As a young, idealistic student matriculating through an honors program at Indiana University, Bloomington, Farina's study bothered my blossoming sensibilities. I generated a number of criticisms of his work, beginning with the fact that hospitalized patients, of course, are going to be less attractive when compared to the general population. It is a hospital, after all, not a beauty salon.

I set about to take a closer look at this issue. Like Farina, I studied patients in a large mental hospital. In my design, and unlike Dr. Farina, I removed the superficial attractiveness visual cues in my subjects' photographs, like hair, makeup and clothes. I also chronicled the severity of the psychiatric diagnoses given my test subjects. I wanted to know whether or not a relationship existed between the severity of the subject's psychiatric diagnosis and level of physical attractiveness.

I identified another question I wanted to answer. Which came first, the chicken or the egg? In other words, did mental illness have a deleterious effect upon looks, and/or was there something about a person's looks (phenotype) that could predict whether or not they would develop a mental illness (genotype related) at some time in the future? Short of having a time machine, how could I answer this question?

I decided that the most practical way to answer this question was to go back and examine the high school yearbook photographs of my test subjects. I incorporated a control group comprised of the high school yearbook photos of my subjects' peers of the same sex.

In both the control group and the test subjects' high school yearbook photos, I removed the attractiveness cues, as I had done with the patient's official hospital photographs. If differences in physical attractiveness during high school were present and preceded a formal diagnosis or first consult with a mental illness, then my study should pick it up.

As part of my study, I established something called independent inter-rater reliability among the people I used to judge the attractiveness of my control groups, my high school subjects, and their hospital photos. I wanted people from both genders and all ages who independently agreed on what a "5" was, for example.

I took my attractiveness raters and had them evaluate the attractiveness of a group of students who had nothing to do with the study. Once I had a representative sample of 10 people who all agreed without knowing they agreed (independent) with their fellow raters (inter-rater reliability), I set about to have that group rate my test subjects and control group photos on a 1 to 10 scale, with 1 being very unattractive and 10 being very attractive.

By the way, years later because of some of the work I was doing with visual image, I had the opportunity to judge nearly 100 beauty pageants. The inter-rater reliability among the judges of these various pageants was never established. This meant, for example, that if one judge rated a girl a 5 and another judge rated the same girl an 8, but both agreed that she had the same degree of attractiveness, the fact that the "8" judge did not use the same personal scale as the "5" judge contaminated the scoring.

I've judged pageants with judges who normalized all their scores. For example, these judges tended to move all scores toward the middle of a 1 to 10 scale. Other judges rated almost everyone with a 9 or 10, while others rated some contestants a 10 while rating a girl who was similarly attractive a 2. Try explaining inter-rater reliability to pageant directors sometime and see where it gets you.

I want to add one more thought about the rating of physical attractiveness. Once people agree on what a "5" means, e.g., is that an average-looking person, better than average-looking person, etc., people very much agree on what is and what is not attractive.

The results of my mental illness study were fascinating and somewhat surprising. I discovered that not only were hospitalized patients less attractive than a matched control group of their peers (no surprise here), but also, as their psychiatric diagnosis became more serious, their attractiveness scores dropped a commensurate amount. Now remember, I removed the visual cues usually associated with physical attractiveness, so all that my raters were left to judge was the shape and contour of the face and the emotiveness or lack thereof of my subjects' static gestalt facial presentation.

Especially surprising to me was that when my subjects' high school yearbook photographs were judged—keep in mind that these photographs were taken before any of my test subjects had received their first mental illness consult or diagnosis—they were judged to be less attractive than their fellow students whose photos immediately preceded their yearbook photos. I investigated something else in my study that I didn't include in the publication version, because I had no idea what I had stumbled upon, and I had to keep my study within limits for publication in the Journal of Abnormal Psychology.

My attractiveness raters reported that "there was something not right" about the faces of the people, even when in high school, who would eventually end up in a mental hospital. When I asked my raters what it was, in particular, they saw in the photos that they described what they saw as "something is not right," none of my raters could articulate a specific part of the face or identifiable facial expression that was "not right."

The raters picked up something related to the gestalt of the face, something that made them feel uncomfortable. Women, I found, were more likely to report the "strangeness" element, as they termed it, than were my male raters. Women appeared to be especially attuned to high school yearbook photos of those boys, not so much girls, who would eventually be hospitalized. When I asked my raters to elaborate about what they were seeing, they told me "something in the eyes wasn't right" about the male hospital patients BEFORE they were ever diagnosed as suffering from a mental illness. Once again, my study had inadvertently confirmed that eyes carry a disproportionately large amount of data about the emotional condition of the person behind the eyes.

My female raters had picked up some visual cue or cues in the phenotypes of the test subjects that made them feel, then think, that something wasn't "right" about my test subjects' high school student genotypes. If this is true, some people may be able to ascertain the biological underpinnings of mental illness before those problems manifest. Whatever it was that some of my female raters picked up, it appeared to be related to *some* women's uncanny ability to ascertain ever-so-subtle visual cues about the absence of reproductive viability and, perhaps, dangerousness in certain males.

Most of us are conditioned to not *consciously* judge others by their looks, and certainly not listen to that little voice in our head that, perhaps, tries to warn us that maybe this person is not a suitable mating choice, or may be dangerous. I think what I stumbled upon goes beyond this rather simple explanation.

I got the strong sense that some women attuned to that little voice in their head could actually pick up potential problems in the genotype of a person destined to become severely mentally ill, as revealed by something in the way future mental patients looked BEFORE they became ill.

Time

Physical attractiveness and time are blood relatives who have been together since their birth. Their relationship is adversarial in nature, in that time consumes beauty and forces the owner of beauty to reevaluate the meaning given to beauty.

Time's effects upon beauty are much more nuanced and subtle than merely its role as the thief in the night that steals beauty. Let me give you a few examples of what I'm talking about.

Human beings tend to get used to repeated sensory stimuli. This is called neuro-habituation. Whether it is a dripping faucet, a squeak in a wooden floor or a beautiful face, we have an initial response that is tempered over time, because our brains habituate to the same repeated or continual sensory stimulation.

What this means is that with the passage of time, beautiful faces seen daily lose some of their initial allure. Conversely, time also works its magic on unattractive faces. Over time, people who may have initially struck us as unattractive become better looking. In both instances, people habituate to visual cues of beauty or unattractiveness, which means that our emotional reaction to beauty or unattractiveness tends to regress to the mean over time.

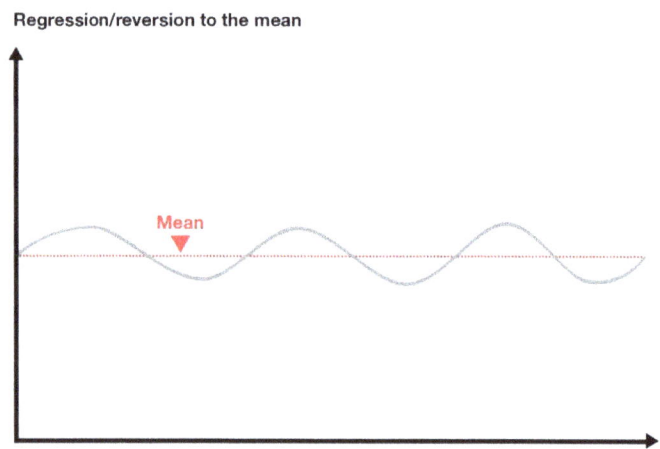

Regression/reversion to the mean

Mean

Above the Mean is Beautiful and Below the Mean is Unattractive

When people are young, time is more of a theoretical concept that can be intellectualized quite easily. This is because the ravages of time don't take effect until one has become a victim of time. Until that point in time, youth tends to delude its owner that the current state of almost everything, including beauty, is permanent. Regardless of what people think or feel about the transient nature of beauty, young women in particular behave as though they will look beautiful forever.

Youthful beauty tends to be experienced as a dichotomous variable; that is, a "yes or no" variable. You're either beautiful or you are not. Over time, however, beauty transforms from a "yes or no" variable into an "it's all relative" variable. "You look good for your age" replaces "you're beautiful."

Time forces the owner of beauty to seek rehab for her likely addiction to beauty. Addiction to beauty is a virtual certainty when a woman is given great beauty as a child or young woman. She becomes addicted to beauty through no fault of her own, but by virtue of the fact that almost any beautiful woman, by virtue of having won the genetic lottery, is elevated to the role of princess—or so she thinks.

Beautiful girls are lavished with gifts and attention beginning in kindergarten. She discovers that she can have her pick of the boys, and soon learns that she can afford to be picky, selective, and play hard to get.

By the time the beautiful girl reaches puberty, she has most likely become very fond of the drug that has served as a tailwind since her first day in preschool. Puberty ushers in changes that result in a full-blown addiction to beauty's benefits, along with a free-floating anxiety about all the envy, jealousy and backstabbing that accompanies the benefits.

Every day, without the beautiful girl's awareness, she is injected with an intoxicating and very addictive drug that becomes an integral part of her personality. The mere thought of withdrawal from this drug is too painful to contemplate. The passage of time, however, forces the now formerly beautiful woman to do just that—withdraw from this most addicting drug.

How addictive is beauty? Some women never make it out of rehab, choosing instead to double down on their addiction to beauty. These women tend to become plastic surgery junkies, because it is plastic surgery that promises to keep her mainlining beauty's intoxicating effects.

There is a point in time when every beautiful woman notices a younger beautiful woman stealing attention from her, in exactly the same way she stole attention from older women when she was young.

As time steals parts of the beauty equation, the void left is replaced with substitutes. The aging beauty that has parlayed her beauty into wealth often substitutes social status for the loss of absolute beauty. Some women become socialites and consumers of expensive clothes, jewelry, cars or other artifacts of material wealth. If this genre of aging beauty is married, her husband's money often becomes the methadone for the former heroin addict.

Some aging women replace the loss of their beauty with a renewed emphasis upon philanthropy. One of the most intriguing substitutions is animal welfare. Animal welfare is not an uncommon life's passion for women who were once stunning beauties and who successfully made it out of beauty rehab. Brigitte Bardot, Doris Day, the Barbie Twins, Pam Anderson, Victoria Beckham, Christie Turlington, Miranda Kerr, Charlize Theron, Anne Hathaway, Isabelle Lucas, and a bevy of other beautiful women have traded in or mitigated their addiction to beauty to work on behalf of innocent creatures that have no concept of physical beauty.

The number of models or movie stars who work on behalf of homeless animals defies the laws of probability. It is not a coincidence that one of the extremely attractive features of dogs, cats, horses and other creatures is that none of them care one bit about how you look, whether you are beautiful or not, whether you are coiffed to perfection or dressed like a street beggar. Is it any wonder that beautiful women and handsome men, who are also kind, end up taking care of and supporting homeless or down-on-their-luck dogs and cats?

Everyone Can Look Better

Readers who are familiar with my various writings understand that I have developed multiple areas of expertise. The first area of expertise I developed involved presentation analysis and modification. What is that?

Presentation of self involves how you look, sound and move, along with the substance or content you communicate. Consistent with the reasoning behind so-called method acting, presentation coaching includes managing your attitudes.

In my work as a clinical psychologist, I found that many psychological problems involving communication and interaction with others could be significantly helped if I modified the patient's presentation dynamics.

Traditionally, clinical psychology, and to a lesser extent psychiatry, modify internal processes, e.g., attitudes, perceptions and cognitions. Psychiatry tends to emphasize psychotropic medications for such presentation complaints as social phobias or debilitating fear of public speaking. I soon realized that modifying a patient's internal state without modifying their presentation dynamics is like clapping with one hand.

One fellow I recall came to me for help because of my growing reputation as a presentation specialist in a totally unrelated field. Later, when he was interviewed by the L.A. Times, he confessed that he had been to several clinical psychologists and psychiatrists, but that none of their interventions made any difference when it came to helping him do better at meeting and getting along with women—that is, until he met me. I've worked with models, actresses, actors, politicians, lawyers, judges and others in the government and military in order to modify their presentations.

How we present ourselves is a matter of chance, not choice. What do I mean by chance? The people you grew up with, i.e., parents, brothers, sisters, and other close friends and relatives you met by chance, significantly influenced how you walk, talk and interact with others. It is fascinating to think that all of the people who influenced your presentation style picked up their presentation dynamics in exactly the same way you did—by chance.

Children born in Paris learn to speak French and, of course, develop a French accent. If you grew up on a Texas ranch, you'll probably walk, talk and act like a Texan rancher. If your father walked a particular way, you will, more likely than not, walk like him.

Unless your mother or father was a professional actor, you probably learned to present yourself through osmosis from a person who never once took the time to analyze their presentation dynamics. For those of you who are not pleased with your presentation dynamics, here is some good news. The fact of the matter is:

Every person, no matter who you are, can improve your level of attractiveness and presentation skills.

In a surprisingly high number of cases, attractiveness can be so dramatically improved that you can jump categories, i.e., move from the unattractive category into the attractive category. Let's now talk a little bit about the specifics of how almost anyone can improve his or her presentation dynamics.

Hair and skin are two areas that can be significantly improved. A minor improvement in hair or skin can result in a tremendous overall increase in one's beauty. Don't think that you have to spend lots of money on haircuts, coloring, or shampoos and conditioners to achieve a marked improvement in your hair. In today's world you can find low-cost shampoos and conditioners online and in grocery and drug stores that rival the expensive hair care products sold only in salons. You can find stylists who will give you a beautiful cut for a discount price, and when it comes to color, the technology has so improved that you can find realistic and beautiful coloring kits on almost any drugstore shelf that at one time were limited to only high-priced salons in Beverly Hills.

I want to emphasize skin care, because a little improvement can make a world of difference. Skin care begins with internal health. Eating well can be the one change you make on the inside that will make you almost immediately look better on the outside. When it comes to skin care, just as with hair care, today's consumer can easily find skin care products that can help transition troubled skin into more beautiful skin in just a few short weeks. Remember, you shed and replace your existing outer layer of skin every 28 days.

For those of you who aspire to have exquisite skin, today's laser technologies are capable of doing what was impossible just a decade ago. Broken blood vessels (telangiectasia) are no match for lasers. Brown spots, fine lines and wrinkles, blotchy and uneven skin can be transformed into beautiful skin in a few short sessions using nothing but light and a good skin care regimen. Stubborn acne can now be effectively treated with medications and/or light therapies that are very effective. For example, take a look at what can be achieved in just ONE treatment with a PICO Genesis laser manufactured by Cutera. Thirty minutes or so, and you can immediately go about your regular routine. And by the way, no pain is involved.

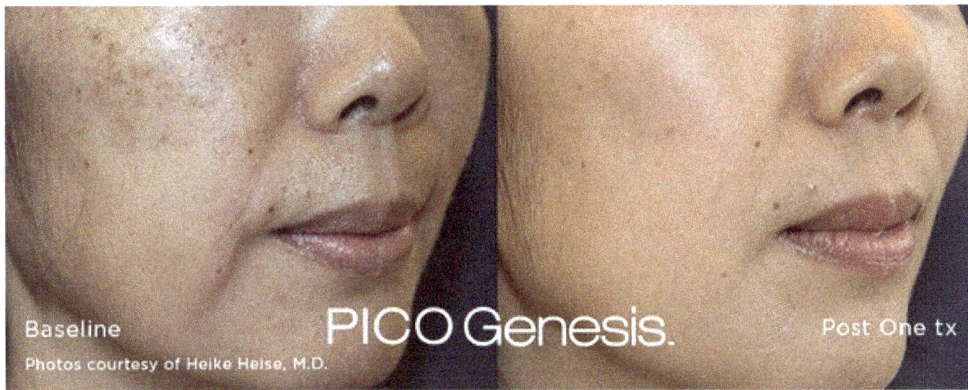

PICO Genesis Laser (Cutera) Results After One Treatment (Photo Courtesy of Dr. Heise, M.D.)

Not only does the patient have noticeably more attractive skin, but also, her confidence level exponentially increased. Self-conscious people concerned about their skin now have reason to hope.

Teeth are yet another area of the body that can increase your overall attractiveness, though you may have to seek the services of a cosmetic dentist to achieve your ideal smile. Dental hygiene will always be your responsibility, no matter who you are or how good your dentist may be. Many people can improve their smile, and thus, their attractiveness, by simply sticking with a good dental hygiene routine. A few good dental hygiene products and some effort, and you will be well on your way to increasing your attractiveness.

If I were to tell you the number of famous people who fail to do the next thing that is fundamental to increasing your attractiveness, you would be shocked: poor overall hygiene.

Years ago I was consulted by an actress and her fiancé for a pre-operative analysis of a cosmetic surgery procedures the client was considering. When the star took a restroom break, her fiancé whispered to me, "Hey doc, could you persuade her to take a shower, brush her teeth and use deodorant more often?"

Poor hygiene is a big deal because it is a much greater problem than anyone realizes. In part, because as I just illustrated, even those people close to us are reluctant to say anything. That's why I am broaching this topic with my readers.

You should talk to yourself about your own personal hygiene and realize that there is no excuse, nor is there any room for error, when it comes to your personal hygiene. And if you are wondering, I did talk to the movie star about her personal hygiene, but not directly. I told her that I had a friend (which I did) in the advertising business who had mentioned to me that he would love to have someone like her do a deodorant ad (also true). The star immediately volunteered that she didn't use deodorant. And without missing a beat, I said to her, "Yes, I know." Silence. I might add this factoid to my story: this situation with the actress occurred during the "grunge period" in Hollywood (1980s). Thank goodness it did not last.

About a year later, she and her fiancé had gone their separate ways, but not because of her personal hygiene problem. I saw her a few times after the breakup, and each time she showed up, the first thing she said was, "Doctor, not only did I put on deodorant, I took a shower just for you." I told her, "Both me and your fans thank you." I won't mention her name, but with a little CSI/cold case investigatory skills, you may be able to guess her name.

I began this book with the promise that you would learn that beauty is a complex and fascinating subject. I've only scratched the surface, even though many of you have concluded that I must have a hidden financial interest in dictionaries and Wikipedia, and enjoy frustrating my readers by demanding so much of them. If you don't already, if you are bright, you will come to appreciate that as my reader, I demand a lot of you. To say that another way, I refuse to talk down to you.

I hope you never look at yourself in the mirror, or anyone else for that matter, the same way. Some of the chapters herein, I hope, made you laugh, think, ponder and feel good. I know from past experiences in teaching this material that a lot of it likely made you feel uneasy. Maybe a part of that uneasiness came from the fact that now you know there are people like me out there who can, simply by looking at you, ascertain your biological, psychological and cultural secrets.

You can take solace in the fact that there aren't that many of "us" out there. And when I am out and about, the last thing I want to do is to exercise my skill set when it comes to visual cue analysis.

This book, I trust, has made you think differently about the package you live in that is called your body. Are you different than the package you live in? When you see a pretty girl or boy, are you seeing what you think you are looking at, or seeing an illusion? I'll leave you with one final story about beauty.

This story is for all of the reactionary social justice warriors who, after reading the introduction to this book, couldn't wait to attack the author over the fact that he failed to take into consideration that some people are blind. "How dare you write about visual images when blind people can't see?"

One of comedian Jerry Seinfeld's classic standup routines includes the following remarks, and I paraphrase: "A person's looks don't impress anyone who's blind, but longer lasting, more substantial qualities do. If you ladies in the audience want to hit on a blind man, you'd better have something to say, and it better make sense."

Seinfeld's satire on the subject of beauty forces one to ponder the fact that blind people, after they come to know you, almost always want to "feel" your face so that they can create a mental image of how you look. Also, one more thing, and this is the thought I want to leave you with: Evidence suggests that wealthy blind men often have beautiful girlfriends and wives.

THE END

INDEX